Revise Chemistry for OCR

Helen Eccles and Mike Wooster

Contents

Introduction – How to use this revision guide

This revision guide is for the OCR Chemistry AS course. It is divided into modules to match the Specification. You may be taking a test at the end of each module, or you may take all of the tests at the end of the course. The content is exactly the same.

Each module begins with an **introduction**, which summarises the content. It also reminds you of the topics from your GCSE course which the module draws on.

The content of each module is presented in **blocks**, to help you divide up your study into manageable chunks. Each block is dealt with in several spreads. These do the following:

- they **summarise** the content;
- they indicate **points to note**;
- they include **worked examples** of calculations;
- they include **diagrams** of the sort you might need to reproduce in tests;
- they provide **quick check** questions, to help you test your understanding.

At the end of each module, there are longer **end-of-module questions** similar in style to those you will encounter in tests. **Answers** to all questions are provided at the end of the book.

You need to understand the **scheme of assessment** for your course.

AS

module name	whole or half module?	what sort of chemistry?	length of exam	total number of marks in the exam	distribution of marks in the exam
2811 Foundation Chemistry	whole module	every branch	1 hour 30 minutes	90 (30%)	*section A* (structured questions) 75 marks *section B* (extended questions) 15 marks
2812 Chains and Rings	whole module	organic	1 hour 30 minutes	90 (30%)	*section A* (structured questions) 75 marks *section B* (extended questions) 15 marks
2813/01 How Far, How Fast?	half module	physical	1 hour	60 (20%)	*section A* (structured questions) 50 marks *section B* (extended questions) 10 marks
Practical chemistry	half module	not covered in this revision guide – your school or college will organise this assessment which makes up 20% of the marks			

OCR AS Chemistry

What will you study to get this qualification?

In the AS year you will complete three modules, one of which is split into two half modules (see table on previous page). This means that you will take three written exams, called "Foundation", "Chains and Rings" and "How Far, How Fast?". There is no synoptic element to the AS year of study – this means each module is self-contained. However, the material in the Foundation module is the basis for the other modules, so you can't forget it just because you've taken the exam.

In the Chains and Rings module there is a "Background Facts" section for the main groups of compounds which gives you an overview of the main concepts and facts. This will point you to the important areas in the section, and give you a useful page for comparison between organic families of compounds for your revision. This is particularly important in the Chains and Rings module because many students find sorting out the differences between organic compounds confusing.

If you are interested in studying chemistry at A2 level, you will find that the OCR A2 course consists of a further three modules; this time two modules are split into half modules.

Good luck with your revision!

Significant figures in calculations

Sig. figs., as they are known, can be important in chemical calculations. How do you work out the number of sig. figs. in a number? The general rule is that *all digits are significant, except zeros that are used to position the decimal point*. Here are some rules to help you apply this generalisation.

- All non-zero digits are significant. So 45.9 has 3 sig. figs., 18.224 has 5 sig. figs.
- Zeros between sig. figs. are significant. So 601.4 has 4 sig. figs., 100.05 has 5 sig. figs.
- Zeros to the left of the first non-zero digit *are not* significant.

So 0.074 has 2 sig. figs., 0.0201 has 3 sig. figs.

- A zero that ends a number and is to the right of the decimal point *is* significant.

So 43.790 has 5 sig. figs., 3.020 has 4 sig. figs.

- Exponentials are used to indicate the number of sig. figs. where a zero ends the number and is to the left of the decimal point. So 5.300×10^3 has 4 sig. figs., 5.3×10^3 has 2 sig. figs.

To work out how many sig. figs. you should give your answer in, look at the figures given in the question. You should use the same number of sig. figs. in your answer as the *lowest* number of sig. figs. given in the question. When you are doing the calculation, use 1 or 2 more sig. figs. than this, and then round off your final answer to the correct number of sig. figs.

Module A: Foundation Chemistry

This is the first module of the AS and A2 course and it is important for two reasons: it contains all the material which is needed to understand the other modules, and topics covered in this module crop up in questions for the other modules.

You will find that some of the material in this module is familiar to you from your GCSE course. This is because the Foundation Chemistry module is a bridge into AS chemistry from GCSE science/chemistry. So, although this module looks very large, don't worry – you know some of it already!

Topic	Reference to specification	Previous knowledge required
Revision of atomic structure, atomic and molecular masses, isotopes and equations	5.1.1, 5.1.2	GCSE material on atomic structure and atomic mass. Constructing and balancing symbol equations.
The determination of atomic masses using the mass spectrometer	5.1.1	Atomic structure and atomic mass.
Detailed electronic arrangements in atoms	5.1.2	GCSE material on the arrangement of electrons in shells.
Ionic, covalent and dative covalent bonding	5.1.3	Ionic and covalent bonding, "Dot-and-cross" diagrams. How atoms become ions.
Intermolecular forces – hydrogen bonds and van der Waals' forces	5.1.3	You may have done some work in GCSE on van der Waals' forces.
The shapes of molecules	5.1.3	The tetrahedral structure of methane from GCSE.
The Periodic Table and periodicity	5.1.4	GCSE work on the structure of the Periodic Table and the locations of Groups and Periods.
The Group 2 elements and their compounds	5.1.5	Knowledge of magnesium, calcium and their compounds.
The Group 7 elements and their compounds	5.1.6	Knowledge of chlorine, bromine and their compounds.

Atoms, molecules and stoichiometry and Atomic structure

The first two sections are covered in a different order to the specification, to make understanding easier. The sections on *ionisation energy* are covered later, in *The Periodic Table: Introduction*.

Quick revision checklist

Make sure you know:

- The relative masses and charges of the proton, neutron and electron

- Atoms are neutral, positive ions (cations) have lost electrons, negative ions (anions) have gained electrons

- **Atomic number Z** = number of protons = number of electrons. This is also the position of the element in the Periodic Table, so for instance element 20 (calcium) has 20 electrons and 20 protons.

- Atomic mass A = no. of protons + no. of neutrons. So A–Z = number of neutrons.

✓ *Quick check 1*

$$^{12}_{6}C$$

- The atomic mass A and atomic number Z are shown as superscripts and subscripts next to the symbol for the element.

- Different **isotopes** of the same element have the same number of protons and electrons but different numbers of neutrons.

- Symbols for common ions (learn these thoroughly!) They are on the next page.

✓ *Quick check 2*

- Symbol equations are important in chemistry because they tell us in what proportions substances react together. Equations have to be **balanced** – they must have the same number of atoms on each side of the arrow. Balance an equation by putting numbers *in front of the formulae only*.

Worked example

Balance this equation: $H_2S + O_2 \rightarrow H_2O + SO_2$
Step 1:   Don't change any of the formulae, just put numbers in front of the formulae until there are the same numbers of atoms on each side of the arrow.
Step 2:   This gives $2H_2S + 3O_2 \rightarrow 2H_2O + 2SO_2$

✓ *Quick check 3*

Constructing equations

Equations can also be *constructed*. This means that if the reactants and products of a reaction are known then the balanced chemical equation can be written down.

Worked example

Sodium reacts with oxygen to give sodium oxide
Step 1:   Think of the word equation: **sodium + oxygen → sodium oxide**
Step 2:   Write down the symbols for the reactants and products: $Na + O_2 \rightarrow Na_2O$
Step 3:   Balance the equation: $4Na(s) + O_2(g) \rightarrow 2Na_2O(s)$

✓ *Quick check 4*

Ionic equations

Equations can be *ionic*.

Ionic equations show just the ions that take part in the reaction.

Worked example

Write down the ionic equation for: **NaOH (aq) + HCl (aq) $\rightarrow$ NaCl (aq) + H$_2$O (l)**

Step 1: Split each ionic compound up into its ions. Leave the covalent molecules as they are.

$$\text{Na}^+(aq) + \text{OH}^-(aq) + \text{H}^+(aq) + \text{Cl}^-(aq) \rightarrow \text{Na}^+(aq) + \text{Cl}^-(aq) + \text{H}_2\text{O}(l)$$

Step 2: Cross out any ions which appear on each side of the arrow (check you are crossing out the same number of each type of ion):

$$\text{Na}^+(aq) + \text{OH}^-(aq) + \text{H}^+(aq) + \cancel{\text{Cl}^-}(aq) \rightarrow \cancel{\text{Na}^+}(aq) + \cancel{\text{Cl}^-}(aq) + \text{H}_2\text{O}(l)$$

Step 3: This leaves the ionic equation: **$\text{OH}^-(aq) + \text{H}^+(aq) \rightarrow \text{H}_2\text{O}(l)$**

Relative Masses

Learn these definitions:

Relative Isotopic Mass
The relative isotopic mass is the mass of an isotope of an element relative to the mass of an atom of ^{12}C (one atom of ^{12}C is given a relative atomic mass of exactly 12).

Relative Atomic Mass A$_r$
The relative atomic mass of an element is the **average mass** of the **naturally-occurring isotopes** of the element relative to the mass of an atom of ^{12}C (one atom of ^{12}C is given a relative atomic mass of exactly 12).

Relative Molecular Mass M$_r$
The relative molecular mass of a compound is the mass of a **molecule** of the compound relative to the mass of an atom of ^{12}C (one atom of ^{12}C is given a relative atomic mass of exactly 12).
Find the relative molecular mass by adding together the relative atomic masses of all the atoms in the molecule. M$_r$ (CH$_3$COOH) = $(12 \times 2) + (16 \times 2) + (1 \times 4) = 60$

Relative Formula Mass M$_r$
The relative formula mass is the relative mass of *one formula unit* of an *ionic compound* relative to the mass of an atom of ^{12}C (one atom of ^{12}C is given a relative atomic mass of exactly 12). The relative formula mass has the same symbol as the relative molecular mass.
M$_r$(MgCl$_2$) = $24 + (35.5 \times 2) = 95$

Cations

sodium	Na$^+$
lithium	Li$^+$
potassium	K$^+$
ammonium	NH$_4^+$
magnesium	Mg^{2+}
calcium	Ca^{2+}
barium	Ba^{2+}
silver	Ag$^+$
iron (II)	Fe^{2+}
iron (III)	Fe^{3+}
lead	Pb^{2+}
copper (II)	Cu^{2+}

Anions

hydroxide	OH$^-$
nitrate	NO$_3^-$
fluoride	F$^-$
chloride	Cl$^-$
bromide	Br$^-$
iodide	I$^-$
sulphate	SO$_4^{2-}$
sulphite	SO$_3^{2-}$
oxide	O^{2-}
carbonate	CO$_3^{2-}$
sulphide	S^{2-}
phosphate	PO$_4^{3-}$

✓ *Quick check 5*

✓ *Quick check 6*

❓ *Quick check questions*

1 How many protons, neutrons and electrons are there in (a) $^{40}_{20}$Ca; (b) $^{16}_{8}$O; (c) $^{14}_{6}$Ca; (d) $^{12}_{6}$C; (e) $^{19}_{9}$F$^-$; (f) $^{27}_{13}$Al^{3+}?

2 Write down the formulae of calcium hydroxide, lead phosphate, barium carbonate and potassium sulphate.

3 Balance these equations:

(a) N$_2$ + H$_2$ $\rightarrow$ NH$_3$ (b) Fe + H$_2$O $\rightarrow$ Fe$_3$O$_4$ + H$_2$

(c) Na$_2$O + H$_2$O $\rightarrow$ NaOH (d) PCl$_5$ + H$_2$O $\rightarrow$ H$_3$PO$_4$ + HCl

4 (a) Construct an equation showing the reaction between sulphuric acid and magnesium, which gives magnesium sulphate and hydrogen as the products.
(b) Give the ionic equation for this reaction.

5 Calculate the relative molecular mass of ethanoic acid, C$_2$H$_4$O.

6 Calculate the relative formula mass of ammonium sulphate.

The mass spectrometer

This instrument measures **atomic masses**. It does this by separating out the different isotopes of an element, and then measuring each one exactly.

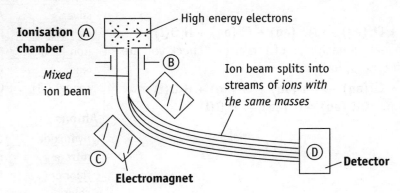

A ionisation – *need to have gaseous ions*

- the element is *vaporised* and passes into this **ionisation chamber**
- here it is bombarded with **high-energy electrons**
- this makes the atoms *lose electrons* and become **ions**

B acceleration - *need the ions to be moving at the same speed*

- a potential difference accelerates the ions
- they all leave the ionisation chamber with approximately the same velocities
- this **ion stream** contains ions of **different isotopic masses**

C deflection – *need the ions to be separated into separate isotopes*

- the ion stream enters a *powerful magnetic field*
- the ions are *deflected* by the field
- lighter ions are deflected more than heavier ions
- this sorts out the mixed ion stream into streams of ions with single isotopic masses

D detection – *need to detect the mass and relative amount of each isotope*

- the magnetic field is increased gradually
- this means that each stream of isotopes hits the detector one at a time (the other isotopes just hit the walls of the instrument)
- the detector converts the information into a mass spectrum

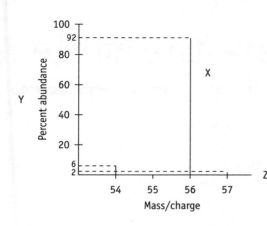

If a sample of iron, Fe, was analysed in the mass spectrometer, the result would look like this – it is called a **mass spectrum:**

X   Each line represents one isotope

Y   This shows the number of atoms of the isotope in every 100 atoms

Z   This shows the mass/charge ratio – but as the charge is +1 it also shows the *mass of the isotope relative to* ^{12}C

The table shows the information the mass spectrum gives you.

isotope	relative mass	relative abundance
1	54	6
2	56	92
3	57	2

This information can be written in the following way, giving the mass of the isotope first with its relative abundance in brackets: 54(6%), 56(92%), 57(2%)

✓ *Quick check 1*

How to calculate the relative atomic mass A_r from the mass spectrum

Step 1:   Multiply the isotopic mass by the % relative abundance:

isotopic mass	% relative abundance	mass of isotopes in 100 atoms
54	6	$54 \times 6 = 324$
56	92	$56 \times 92 = 5152$
57	2	$57 \times 2 = 114$

Step 2:   Add all these together: **324 + 5152 + 114 = 5590**

Step 3:   Divide by 100: $A_r = \dfrac{5590}{100} - 55.90$

So the relative atomic mass of the sample is 55.90 (no units).

✓ *Quick check 2*

> ❶ Mass spectrometry is used to determine whether water supplies are contaminated with pollutants, and to detect performance-enhancing drugs in the blood of athletes and animals.

> ❶ One of the most recent uses of mass spectrometry was on the 1999 Mars space probe. The mass spectrometer is a small instrument so it could conveniently be put onto the probe, and at Mars it analysed the atmosphere to determine its composition.

?   *Quick check questions*

1   Sketch the mass spectrum of magnesium which has isotopes 24(79%) 25(10%) 26(11%).

2   Calculate the relative atomic masses of the following elements. (Give your answer to four significant figures.):

(a)   silver   107(51.3%) 109(48.7%)

(b)   chlorine   35(75.5%) 37(24.5%)

Amount of substance - the mole

Equations tell us the proportions in which substances react together. We need to be able to convert those amounts into grams, so we use the **mole**, which is a particular amount of substance. A mole of any one substance has the **same number of particles** as a mole of another substance. Because atoms of different substances have different masses, a mole of any one substance has a **different mass** to a mole of another substance.

> **A mole** of a substance is the **amount of substance** that contains **6 x 10²³ particles**, which is the same number of particles as there are atoms in exactly 12 g of ^{12}C.
>
> The number 6×10^{23} is called the **Avogadro constant or Avogadro number,** symbol **L**, units **mol⁻¹**.

✓ *Quick check 1*

Chemists often work out the mass in g of 1 mole of substance – this is called the **molar mass**

> **Molar mass, M**
>
> The molar mass is the mass of 1 mole of the substance in g.
>
> It has the symbol M and units of g mol⁻¹.
>
> It is simply the M_r value with units!

▶ The mole refers to particles – these can be atoms, molecules or ions.

✓ *Quick check 2*

Worked example

Find the molar mass of sulphuric acid, **H₂SO₄**. **A_r (H) = 1; A_r (S) = 32; A_r (O) = 16.**

Work this out exactly as for a relative molecular mass, but add units.

Step 1: Add up the relative masses of all the atoms:

$$(1 \times 2) + 32 + (16 \times 4) = 98$$

Step 2: Add the units: **98 g mol⁻¹**.

Worked example

Find the formula mass of lithium bromide.

Step 1: Put down the formula of lithium bromide: **Li₂Br**

Step 2: Add up the relative masses of all the atoms

$$A_r(Li) = 7, A_r(Br) = 80: (2 \times 7) + 80 = 94$$

Step 3: Add the units: **94 g mol⁻¹**.

A word about UNITS!
A relative mass has **no units:**

relative atomic mass......no units

relative molecular mass.......no units

relative formula mass......no units

BUT the units for **molar mass M** are **gmol⁻¹** AND the units for the **Avogadro constant L** are **mol⁻¹**

Mole calculations

There are 3 main types of equation involving the mole, for solids, solutions and gases.

For solids:

elements: **number of moles** $= \dfrac{\text{mass in g}}{A_r}$ compounds: **number of moles** $= \dfrac{\text{mass in g}}{M_r}$	Can be rearranged: mass in g = no. moles $\times$ A_r or M_r A_r or $M_r = \dfrac{\text{mass in g}}{\text{no. of moles}}$

Worked example

How many moles are there in 5 g of calcium nitrate, $Ca(NO_3)_2$?

Step 1: Work out the M_r of calcium nitrate:

$Ca(NO_3)_2 = 40 + (14 \times 2) + (16 \times 6) = 164$

Step 2: Apply the equation: **no. of moles** $= \dfrac{\text{mass in g}}{M_r} = \dfrac{5}{164} = 0.030$ **moles**

✓ *Quick check 3*

For solutions:

number of moles = volume in dm^3 $\times$ concentration in mol dm^{-3} *or* **number of moles** $= \dfrac{\text{volume in cm}^3}{1000} \times$ **concentration**	Can be rearranged: vol. in dm$^3 = \dfrac{\text{no. mol}}{\text{concn}}$ (vol. in cm$^3 \times 10^{-3}$) $= \dfrac{\text{no. mol}}{\text{concn}}$ concn $= \dfrac{\text{no. mol}}{\text{vol. in dm}^3}$ concn $= \dfrac{\text{no. mol}}{(\text{vol. in cm}^3 \times 10^{-3})}$

Worked example

How many moles of sodium hydroxide are there in 10 cm^3 of 0.1 mol dm^3 aqueous sodium hydroxide?

Step 1: Apply the equation:

no. of moles $= \dfrac{\text{vol. in cm}^3}{1000} \times$ **concn** $= \dfrac{10}{1000} \times 0.1 = 0.001$ **moles**

For gases:

number of moles $= \dfrac{\text{volume in dm}^3}{24}$ *or* **number of moles** $= \dfrac{\text{volume in cm}^3}{24000}$	Can be rearranged: vol. in dm^3 = no. of moles $\times$ 24 vol. in cm^3 = no. of moles $\times$ 24000

REMEMBER!! 1 mole of gas occupies 24 dm^3 at room temperature and pressure.

Worked example

How many moles of oxygen occupy 600 cm^3 at room temperature and pressure?

Step 1: Apply the equation:

no. of moles $= \dfrac{\text{vol. in cm}^3}{24000} = \dfrac{600}{24000} = 0.025$ **moles**

Types of mole calculation

Mass of a solid from the number of moles

Worked example

What is the mass of 2 moles of calcium sulphate, $CaSO_4$?

Step 1: Calculate the M_r of $CaSO_4$: $M_r = 40 + 32 + (16 \times 4) = 136$

Step 2: Use the rearranged mole equation for solids

$$\text{mass in g} = \text{no. of moles} \times M_r = 2 \times 136 = 272 \text{ g}$$

Mass of a solid from an equation

Worked example

What mass of calcium carbonate reacts completely with 25 cm^3 of 2.0 mol dm^{-3} hydrochloric acid?

Step 1: Construct the equation:

$$CaCO_3(s) + 2HCl(aq) \rightarrow CaCl_2(aq) + CO_2(g) + H_2O(l)$$

Step 2: Calculate the number of moles of hydrochloric acid used in the reaction:

$$\text{no. of moles of HCl} = \frac{\text{vol. in cm}^3}{1000} \times \text{concn} = \frac{25}{1000} \times 2 = 0.05 \text{ mol}$$

Step 3: From the equation see how many moles of calcium carbonate are needed.

1 mole calcium carbonate required by 2 moles hydrochoric acid.

$\therefore$ 1 mole hydrochoric acid requires $1 \div 2$ moles calcium carbonate.

$\therefore$ 0.05 moles hydrochoric acid require: $0.05 \div 2$ moles = 0.025 mol $CaCO_3$

Step 4: Now use the mole equation to find the mass of calcium carbonate

$$\text{mass of CaCO}_3 = \text{no. mol} \times M_r = 0.025 \times [40 + 12 + (16 \times 3)] = 2.5 \text{ g}$$

Using an equation to find the mass of product formed

Worked example

What is the mass of zinc sulphate produced when 4.0 g of zinc is reacted with excess sulphuric acid?

Step 1: Construct the equation for the reaction:

$$Zn(s) + H_2SO_4(aq) \rightarrow ZnSO_4(aq) + H_2(g)$$

Step 2: Calculate the number of moles of zinc used:

$$\text{no. of moles of zinc} = \frac{\text{mass in g}}{A_r} = \frac{4}{65} = 0.06 \text{ mol}$$

Step 3: Look at the equation to see the number of moles of zinc sulphate formed

0.06 moles of Zn gives 0.06 mol of ZnSO₄

Step 4: Calculate the mass of ZnSO₄

mass of ZnSO₄ = no. of moles × Mᵣ = 0.06 × [65 + 32 + (16 × 4)] = 9.7 g.

Using an equation to find the volume of a gas

Worked example

In the worked example above, what volume of hydrogen in cm^3 is produced, at room temperature and pressure?

Step 1: The number of moles of zinc used is already known: 0.06 mol

Step 2: Look at the equation to find the number of moles of hydrogen produced

0.06 moles of Zn gives 0.06 moles of H₂

Step 3: Find the volume from the number of moles

volume in cm^3 = no. of moles × 24000 = 0.06 × 24000 = 1440 cm^3

> REMEMBER!! 1 mole of gas occupies 24 dm³ at room temperature and pressure.

Working out the concentration of a solution

Worked example

Calculate the concentration of 2.80 g of potassium hydroxide, KOH, in 500 cm^3 of solution

Step 1: Calculate the number of moles of solid KOH used

$$\text{no. of moles} = \frac{\text{mass in g}}{M_r} = \frac{2.80}{56} = 0.050 \text{ mol of KOH}$$

Step 2: Now find the concentration of the KOH solution

$$\text{concn} = \frac{\text{no. of moles}}{\text{vol. in } cm^3 \times 10^{-3}} = \frac{0.050}{500 \times 10^{-3}} = 1.00 \text{ mol dm}^{-3}$$

✓ *Quick check 4*

Quick check questions

1 How many atoms are there in 1 mole of silicon atoms?

2 What mass of methane CH₄ contains 12×10^{23} molecules?

3 How many moles of Ag are produced in the decomposition of 20 g of silver oxide? $2Ag_2O(s) \rightarrow 4Ag(s) + O_2(g)$

4 A student measured the amount of ammonia produced in the following reaction: $Ca(OH)_2(s) + 2NH_4Cl(s) \rightarrow CaCl_2(s) + 2NH_3(g) + 2H_2O(g)$

The ammonia had a volume of 600 cm^3 at room temperature and pressure.

(a) What mass of ammonium chloride was used in the experiment?

(b) What was the mass of calcium chloride collected after removal of the water?

Give your answer to two decimal places.

Aᵣ values	
Ag	108
O	16
N	14
H	1
Cl	35.5
Ca	40

Empirical formulae

The **empirical formula** of a compound is the *simplest whole number ratio* of the number of atoms of each element in 1 molecule – the simplest mole ratio.

The **molecular formula** is the *actual number of atoms* of each element in 1 molecule of the substance.

Empirical formulae are calculated from the masses of the elements in the compound. The best way of doing this is to tabulate the data and calculations.

Empirical formula from the mass of elements reacting together

Worked example

Calculate the empirical formula of silicon oxide if 3.5 g of silicon combines with 4.0 g of oxygen.

Step 1: Write down the mass of each element in the table

Step 2: Calculate the number of moles of each element in the table (divide mass by A_r)

Step 3: Divide the number of moles of each element by the smallest number that goes into both. This is the empirical formula.

(If this is not obvious, divide by the smallest number of moles – this automatically gives a ratio of 1 for one element – then multiply by 2, or 3, until a whole number for every element is obtained.)

✓ Quick check 2

	Si	O
mass in g	3.5	4.0
number of moles	$\frac{3.5}{28} = 0.125$	$\frac{4.0}{16} = 0.25$
mole ratio	$\frac{0.125}{0.125} = 1$	$\frac{0.25}{0.125} = 2$
empirical formula	Si_1	O_2
	which we write as SiO_2	

Empirical formula from the mass of one reactant and the mass of product

Worked example

Calculate the empirical formula of iron bromide if 3.78 of iron react with bromine to give 20.0 g of iron bromide.

Step 1: First find the mass of bromine combined with the iron, by subtracting the mass of iron from the mass of product:

mass of bromine = 20.0 – 3.78 = 16.22 g

Step 2: Then work out the mole ratio of the reactants as before.

	Fe	Br
mass in g	3.78	16.22
number of moles	$\dfrac{3.78}{56} = 0.0675$	$\dfrac{16.22}{80} = 0.202$
mole ratio	$\dfrac{0.0675}{0.0675} = 1$	$\dfrac{0.202}{0.0675} = 3$
empirical formula	Fe_1 which we write as **$FeBr_3$**	Br_3

Empirical formula from the percentage composition by mass

Sometimes the percentage of each reactant is given instead of its mass. In this case, assume you have 100 g of the compound. The percentage of the element then becomes its mass in g.

✓ Quick check 1

Worked example

Calculate the empirical formula of a compound consisting of 47.4% sulphur and 52.6% chlorine

Step 1: Assume 100 g of compound so mass of S = 47.4 g and mass of Cl = 52.6 g

Step 2: Continue the calculation as usual

	S	Cl
mass in g	47.7	52.6
number of moles	$\dfrac{47.7}{32} = 1.49$	$\dfrac{52.6}{35.5} = 1.48$
mole ratio	1	1
empirical formula	S_1 which we write as **SCl**	Cl_1

Finding the molecular formula from the empirical formula

If the relative molecular mass is known then the molecular formula can be worked out from the empirical formula.

✓ Quick check 2

Worked example

In the example above, if the relative molecular mass is 135, find the molecular formula.

Step 1: Calculate the relative mass of the empirical formula

Step 2: Let the molecular formula be **(SCl)** $\times n$

Step 3: Then
$$(SCl) \times n = 135$$
$$67.5 \times n = 135$$
$$n = 135/67.5 = 2$$

So the molecular formula is **S_2Cl_2**

❓ Quick check questions

1. Dibutyl succinate is a domestic insect repellent. Its composition is 62.58%C, 9.63%H and 27.79%O. Its relative molecular mass is 230. What are the empirical and molecular formulas of dibutyl succinate?

2. Calculate the empirical formula of the substance which, on analysis, is found to contain 0.42 g C, 0.11 g H and 0.29 g O.

Titration calculations

These often cause problems. One way of solving them successfully is to follow this plan, and set out the calculations underneath each reactant or product in the equation.

✓ *Quick check 1 & 2*

Worked example

A student doing a titration found that 12.5 cm^3 of 0.020 mol dm^3 aqueous KOH exactly neutralised 25.0 cm^3 of nitric acid. What is the concentration of the nitric acid?

Step 1:
Construct the equation: **KOH (aq) + HNO$_3$ (aq) → KNO$_3$ (aq) + H$_2$O (l)**.
You are only interested in the KOH and the HNO$_3$.

	KOH	**HNO$_3$**
Step 2: Write down what you know	vol. = **12.5 cm^3** concn = **0.02 mol dm^{-3}**	vol. = **25 cm^3** concn = **? mol dm^{-3}**
Step 3: Calculate the number of moles of KOH	no. of moles = vol. × concn no. of moles = $\frac{12.5}{1000} \times 0.02$ = **2.5 × 10^{-4} mol**	
Step 4: Work out the number of moles of HNO$_3$ by looking at the equation		no. of moles = **2.5 × 10^{-4} mol** (1:1 stoichiometry in the equation)
Step 5: Work out the concentration of the HNO$_3$		concn = $\frac{\text{no. of moles}}{\text{vol.}}$ concn = $\frac{2.5 \times 10^{-4}}{25 \div 1000}$ = **0.01 mol dm^{-3}**

❓ Quick check questions

1 A sample of 10 cm^3 of 0.07 mol dm^{-3} sulphuric acid required exactly 23 cm^3 of aqueous sodium hydroxide to neutralise it. Calculate the concentration of the aqueous sodium hydroxide.

2 What volume of 0.1 mol dm^{-3} aqueous sodium hydroxide is required to neutralise 25 cm^3 of 0.2 mol dm^{-3} hydrochloric acid?

Electron configurations

Electrons fill different energy levels, or **shells**. These shells can be subdivided further into **sub-shells**, and the sub-shells are made up of **orbitals**. The first shell is nearest the nucleus. The sub-shells are called **s, p** and **d sub-shells**, and each one holds a different number of electrons.

This is how the sub-shells are divided between the shells:

shell	shell 1	shell 2		shell 3		
sub-shell	s	s	p	s	p	d
max no. of electrons in sub-shell	2	2	6	2	6	10

The order in which the sub-shells are filled is from the lowest energy level to the highest, which is

1s 2s 2p 3s 3p 4s 3d

Each **sub-shell** is made up of one or more **orbitals**; each **orbital** contains a maximum of **two electrons** only.

> Make sure you understand the difference between shells, sub-shells and orbitals.

Worked examples

✓ *Quick check 1*

Br has 35 electrons: electron configuration is $1s^2 2s^2 2p^6 3s^2 3p^6 4s^2 3d^{10} 4p^5$
Br^- has 36 electrons: electron configuration is $1s^2 2s^2 2p^6 3s^2 3p^6 4s^2 3d^{10} 4p^6$
Na^+ has 10 electrons (Na has 11 electrons): electron configuration is $1s^2 2s^2 2p^6$

Shapes of the s orbitals and p orbitals

The s and p orbitals have different shapes. The s orbitals are spherical. There are 3 p orbitals at right angles to each other, and they have a lobed shape.

s orbital

p orbitals

Electron configurations and the Periodic Table

Look at the Periodic Table (see page 23). All the elements in a particular Group have the same type of electronic configuration.

✓ *Quick check 2*

Group	1	2	3	4	5	6	7	8
electron configuration (n is the number of the Period)	ns^1	ns^2	$ns^2 np^1$	$ns^2 np^2$	$ns^2 np^3$	$ns^2 np^4$	$ns^2 np^5$	$ns^2 np^6$

This shows us why elements in the same Group have similar chemical properties:

- it is the electrons that are involved when an element reacts
- so elements with electrons in the same type of orbital will react in a similar way.

? Quick check questions

1 Write down the electron configurations of V, K^+, O^{2-}.

2 Which element has the electron configuration $1s^2 2s^2 2p^6 3s^2$?

Chemical bonding

Ionic bonding

- This is the bonding usually found in compounds of a metal (Groups 1 and 2) with a non-metal (usually Groups 5–7).
- Ionic bonding is the **electrostatic attraction between two oppositely-charged ions.**
- It arises when an electron is **transferred** from one atom to another, a process which forms **ions.**

Example

Potassium chloride, KCl, has ionic bonding. It is composed of K^+ and Cl^- ions. These ions are formed from the atoms by the loss or gain of an electron.

The K atom has 1 electron in the outermost 4s sub-shell.

The Cl atom has 7 electrons in its outermost 3s and 3p sub-shells.

The K atom loses an electron to become the potassium ion:

$$K \rightarrow K^+ + e^-$$

and this electron is transferred to the chlorine atom which becomes a chloride ion:

$$Cl + e^- \rightarrow Cl^-.$$

By doing this, both achieve a **full shell of 8 electrons.**

We can show ionic bonding by "dot-and-cross" diagrams – dots indicate electrons of one species, and crosses indicate electrons of the other species. A dot-and-cross diagram of the formation of KCl looks like this:

$$K^{\times} \qquad {}^{\circ}_{\circ}\overset{\circ\circ}{\underset{\circ\circ}{Cl}}{}^{\circ}_{\circ} \longrightarrow \left[K \right]^+ \left[{}^{\circ}_{\times}\overset{\circ\circ}{\underset{\circ\circ}{Cl}}{}^{\circ}_{\circ} \right]^-$$

| potassium atom K | chlorine atom Cl | potassium ion K^+ | chlorine ion Cl^- |

Here are the dot-and-cross diagrams for two ionic structures, sodium chloride and magnesium oxide:

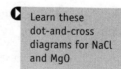

Learn these dot-and-cross diagrams for NaCl and MgO

sodium chloride

$$Na \longrightarrow Na^+ + e^-$$
$$Cl + e^- \longrightarrow Cl^-$$

$$\left[Na \right]^+ \left[{}^{\circ}_{\times}\overset{\circ\circ}{\underset{\circ\circ}{Cl}}{}^{\circ}_{\circ} \right]^-$$

magnesium oxide

$$Mg \longrightarrow Mg^{2+} + 2e^-$$
$$O + 2e^- \longrightarrow O^{2-}$$

$$\left[Mg \right]^{2+} \left[{}^{\times}_{\circ}\overset{\times\circ}{\underset{\times\circ}{O}}{}^{\circ}_{\circ} \right]^{2-}$$

Solid ionic compounds have a **giant lattice structure,** where each ion is next to ions of opposite charge and is held in place by the attraction between them.

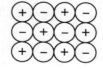

Covalent bonding

- This is the bonding usually found in compounds of a non-metal with another non-metal (usually Groups 4–7).
- Covalent bonding is the **sharing of a pair of electrons** between two atoms. We can show covalent bonding by "dot-and-cross" diagrams.

Example

Chlorine gas, Cl_2, has covalent bonding.

Both Cl atoms have 7 electrons in the outer sub-shells. Two Cl atoms combine so that each shares the single electron of the other. Both chlorine atoms then have a full outer sub-shell of 8 electrons. A covalent bond is represented by a single line, so that chlorine, Cl_2, is drawn as Cl–Cl.

In covalent compounds it is possible for two atoms to share more than one pair of electrons. Sharing two pairs of electrons gives a **double bond**.

Sharing 3 pairs of electrons gives a **triple** bond.

Learn the dot-and-cross diagrams for these covalent compounds					
hydrogen H_2	H–H	H o x H	methane CH_4	H–C–H (with H above and below)	H o x C o x H (with H above and below)
oxygen O_2	O=O	O x O (dot-and-cross)	carbon dioxide CO_2	O=C=O	O x C x O (dot-and-cross)
hydrogen chloride HCl	H–Cl	H x Cl (dot-and-cross)	ethene $H_2C=CH_2$	H–C=C–H (with H's)	H x C x C x H (with H's)

Dative covalent bonding

- This is a *particular type of covalent bonding*.
- In a dative covalent bond, both the electrons in the bond are supplied by **one atom** only.
- Dative covalent bonds are sometimes called **co-ordinate** bonds.

Example: NH_4^+ has dative covalent bonding in one of the bonds between the N and H atoms shown by an arrow.

1 (a) Draw dot-and-cross diagrams of C_2H_6; KBr; H_2O; $MgCl_2$; H_3O^+; N_2

   (b) Describe the type of bonding in each of these compounds.

Polarisation and intermolecular forces

A polar bond is a bond with a **charge separation**. It occurs in covalent molecules, and when it does we say the molecule has a **permanent dipole**.

Polarisation of the bond in covalent molecules

Hydrogen chloride H–Cl

- The δ+ sign and the δ– sign mean the electrons are not evenly distributed between the two atoms, but are concentrated on the Cl atom. This is because Cl attracts electrons more than H.
- We say that chlorine has a greater **electronegativity** than hydrogen.

> **Electronegativity** is the ability of an atom in a covalent bond to attract the bonding electrons.

Bond polarisation means that the covalent bond is becoming ionic in character, because it has a dipole moment.

Chlorine Cl–Cl

In this molecule both atoms have the same electronegativity so there is **no bond polarisation**.

Polarisation of the anion in ionic bonds

Magnesium carbonate is an ionic compound consisting of Mg^{2+} cations and CO_3^{2-} anions in a giant structure. The Mg^{2+} cation is very small and highly charged – it has a **high charge density.** The CO_3^{2-} anion is much, much larger than the Mg^{2+} cation, and has a **diffuse** structure – it is well spread out and has a **low charge density.** This means that the Mg^{2+} cation can get close to the CO_3^{2-} anion and distort it by pulling electrons towards itself – we say the anion has been **polarised.** This weakens the bond between the ions.

> **Summary**
>
> - **Bond polarisation** occurs in **covalent** compounds. It arises because the two atoms sharing a bond have different electronegativities.
> - **Polarisation of the anion** occurs in **ionic** compounds. It arises if the metal cation has a high charge density and the anion has a low charge density.
> - Polarisation shows us that ionic bonds and covalent bonds are the two extremes of bonding types, and there is a gradual range of bonding in between.

δ+ δ–
H—Cl

> Halogen atoms generally have high electronegativity.

> A *bond* is polar if there is a large difference in electronegativity between between the atoms.

✓ *Quick check 1*

> To discover if a *molecule* is polar, you have to know its shape: draw a dot-and-cross diagram (see p. 21).

? *Quick check question*

1 Which of the following molecules are polar? Show where the dipole moment is on the molecule. HI; PCl_3; CH_4.

Intermolecular forces

These are forces which arise *between* molecules. They are **short-range** forces.

- **Permanent dipole - permanent dipole forces** These occur in molecules with a permanent dipole. The δ+ atom on one molecule attracts the δ– atom on a neighbouring molecule, so electrostatic attractions operate between the molecules.

   Examples are H-Cl, H-Br, H-I, SO_2, PH_3.

- **Instantaneous dipole–induced dipole forces** These occur in all atoms and molecules, but are only significant where there are lots of electrons. They are also called **van der Waals' forces.** They are stronger for long, sausage-shaped molecules than small, round molecules.

   Examples are the noble gases, the halogens, long polymer chains. These forces are the weakest type of intermolecular force.

- **Hydrogen bonding** This is a strong type of permanent dipole – permanent dipole attraction that occurs between molecules containing **H** attached to either **N, O,** or **F.** These molecules have a permanent dipole so the δ+ H atoms are attracted to the δ– N, O or F atoms and a hydrogen bond is formed. We represent hydrogen bonds by showing three dashed lines between the atoms concerned. Hydrogen bonds are stronger than permanent dipole – permanent dipole attractions because the N, O, F atoms are highly electronegative.

   Examples of molecules which show hydrogen bonding are NH_3, H_2O, HF.

✓ Quick check 1

These attractions arise because the electron cloud in the atom or molecule is constantly moving. At any instant, there will be a higher density of electrons in one area than in another, and this means an instantaneous dipole has been created. The instantaneous dipole induces a dipole in a neighbouring molecule, and so the two atoms or molecules are attracted to each other.

✓ Quick check 2 & 3

H_2O and hydrogen bonding

Water behaves strangely.... because of its hydrogen bonds! A water molecule can form 2 hydrogen bonds to other water molecules.

- Water has a relatively high boiling point compared to the other hydrides in Group 6 because hydrogen bonds have to be broken.

- Water forms a meniscus because the hydrogen bonds pull the water surface downwards.

- When water freezes into ice, a whole network of hydrogen bonds form. The hydrogen bonds and covalent O-H bonds are arranged tetrahedrally around the O atom. The H_2O molecules are kept apart to form a lattice by this arrangement, and there is lots of space in the structure – this makes ice have a *lower density* than water.

Students often forget that H-bonding occurs **only with N,O,F** – no other atoms!

? Quick check questions

1. Predict the main type of intermolecular force in C_3H_8; NH_2OH; PCl_3.

2. Draw a diagram showing the hydrogen bonds in the alcohol ethanol, C_2H_5OH.

3. Suggest a reason why ethanol has a much higher boiling point than methoxymethane, CH_3-O-CH_3, which has a similar structure.

Structures and physical properties

The main types of structure you must know about are giant ionic lattices, discrete molecular substances, giant covalent lattices and metals.

- Giant ionic lattices have ionic bonding.
- Discrete molecular and giant molecular structures have covalent bonding.
- Metals have metallic bonding.

Giant ionic lattices - Examples are found in NaCl, $MgCl_2$, MgO, Na_2O, NaOH, KBr. - Ionic compounds have a giant lattice structure of positive and negative ions, held in place by electrostatic attraction.	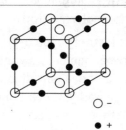 ○ – ● +
Discrete molecular structures - Examples are the diatomic gases such as H_2, Cl_2, N_2, and some solids like P_4. - Discrete molecular substances have individual molecules with intramolecular covalent bonding. There is very little attraction between molecules.	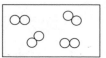 Diatomic gas molecule
Giant molecular structures - These are found in diamond, graphite (within the layers, not between layers), silicon dioxide. - Giant molecular substances have a giant lattice of atoms linked together with covalent bonds.	
Metals - Examples are Cu, Mg, Na, Al. - Metallic structures are composed of a lattice of positive ions in a sea of mobile electrons. This is called **metallic bonding**. It arises because the outermost electron shells overlap each other and merge together. The electrons in the outer shells can travel from atom to atom, and positive ions are left in their lattice position. - Electrostatic attraction between the mobile electrons and the positive ions holds the structure together.	

How the physical properties of a substance are linked to its structure

Ionic structures

Physical properties	Explanation
high melting point	strong attraction between ions
no electrical conductivity when solid	ions are fixed in the lattice
good electrical conductivity when molten	the ions can move, so electricity can flow
no heat conductivity	ions are fixed in position so cannot transfer heat energy vibrations
dissolve well in polar solvents such as water	the polar solvent molecules attract the ions out of the lattice into solution

Discrete molecular structures

Physical properties	Explanation
low melting point	molecules are far apart with little intermolecular forces
no electrical conductivity	no free electrons or ions
no heat conductivity	molecules are too far apart

Giant molecular structures

Physical properties	Explanation
very high melting point	covalent bonds are very strong, so it takes a lot of energy to break them
no electrical conductivity	electrons are fixed in the covalent bonds, so cannot move to carry current
no heat conductivity	atoms are fixed in place by the network of covalent bonds so cannot transfer heat energy vibrations

Metals

Physical properties	Explanation
high melting point (but not as high generally as giant molecular compounds)	electrostatic attraction between ions is strong
good electrical conductivity	electrons are mobile, so are free to carry a current
good heat conductivity	ions are loosely fixed in the mobile sea of electrons, so easily transfer vibrations of heat energy

> **?** *Quick check questions*
>
> 1 Name the type of structure in (a) KNO_3 (b) K (c) Kr (d) CH_4.
>
> 2 Silicon dioxide, SiO_2, is commonly known as sand. From your knowledge of the physical properties of sand, suggest a structure of SiO_2 and explain your answer.

Shapes of molecules - Electron Pair Repulsion Theory

Electron pairs arrange themselves so they are as far apart as possible. In a covalent compound, the number of electron pairs around the central atom determines the shape of the molecule.

2 electron pairs	linear	$BeCl_2$	
			$\overset{\times\times}{\underset{\times\times}{\times}}Cl\overset{\times}{\underset{\times}{\circ}}Be\overset{\times}{\underset{\times}{\circ}}\overset{\times\times}{\underset{\times\times}{Cl}}\times$
			Cl—Be—Cl 180° bond angle 180°
3 electron pairs	trigonal	BF_3	
			$F\ B\ F$ structure with F at top
			120° F—B with F, F bonds bond angle 120°
4 electron pairs	tetrahedral	CH_4 NH_4^+	
			$H\overset{\times}{\underset{\circ}{}}C\overset{\times}{\underset{\circ}{}}H$ with H above and below
			109.5° tetrahedral CH₄ structure bond angle 109.5°

We know the different shapes that molecules have by working out how the electron pairs making up the bonds arrange themselves in space. A set of rules called **electron pair repulsion theory (EPRT)** tell us that:

● electron pairs arrange themselves so they are as far apart as possible

● an electron pair shared between two atoms is called a **bonding pair (BP)**

● an electron pair on one atom only – not shared – is called a **lone pair (LP)**

● **LP-LP repulsion > LP-BP repulsion > BP-BP repulsion**

These rules explain the different shapes of the ammonia, NH_3, water, H_2O and carbon dioxide, CO_2 molecules.

NH₃

The nitrogen atom has four pairs of electrons, so these will take up a basically tetrahedral shape. But one electron pair is a lone pair, so it repels the shared pairs more than a bonding pair would and the bond angle is decreased. The final shape of the molecule is called **pyramidal.**

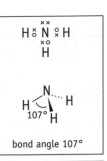

bond angle 107°

H₂O

The oxygen atom has four pairs of electrons, so these will take up a basically tetrahedral shape. But two electron pairs are lone pairs, so the shared pairs are repelled more than in NH_3, and the bond angle in H_2O is less than in NH_3. The final shape of the molecule is called **bent,** or **non-linear.**

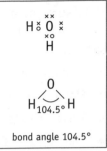

bond angle 104.5°

What about multiple bonds?

If a molecule has double (or triple) bonds, the bonding electron pairs are located between the bonding atoms, just like single bonds, so the same rules apply. Look at carbon dioxide:

◖ Don't confuse the shape of the molecule with the shape the electron pairs adopt.

CO₂

A dot-and-cross diagram of CO_2 shows that the bonding is O=C=O. The electron pairs in the double bonds will repel each other as much as possible so the final shape of the molecule is **linear.**

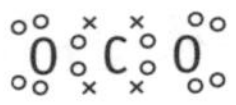

What about a non-linear molecule with double bonds? A good example of this is sulphur dioxide:

SO₂

Here the dot-and-cross diagram shows double bonds between the S and O atoms, just like CO_2, but there is also a lone pair on the sulphur atom. The lone pair repels the two double bonds, and the molecule is a **bent** or **non-linear** shape.

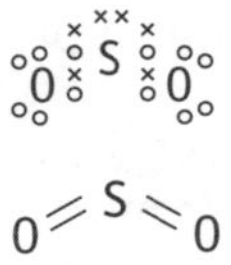

Worked example

To work out the shape of a molecule, for example H_3O^+

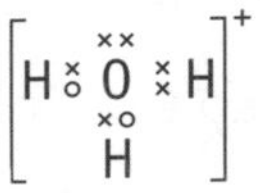

Step 1: Draw a dot-and-cross diagram of the molecule.

Step 2: See what shape the electron pairs will adopt. They will be arranged tetrahedrally.

Step 3: Now see if there are any lone pairs. Yes, one lone pair.

Step 4: EPRT tells you that the lone pair will repel the bonding pairs and the molecule should have the same shape and bond angle as NH_3.

H_3O^+ has a pyramidal shape with a bond angle of 107°.

? *Quick check question*

1 Sketch the shapes and predict the bond angles in the molecules ClF; PCl_3; OCl_2; XeF_6; CCl_4.

The Periodic Table

The Periodic Table of the Elements comes in various shapes and forms, but the one you will use in the examinations is shown on page 23.

Make sure you are familiar with the names and positions of the main **Groups** (elements in the same column) and the numbering of the **Periods** (elements in the same row).

- Elements in the same Group have *similar* physical and chemical properties.
- Elements in the same Period have *repeating* physical and chemical properties.
- The elements are arranged by **increasing atomic number Z**.
- The Periodic Table can be divided into **blocks**, according to the electronic configurations of the elements.

The **s-block** contains those elements with outer electrons in the s sub-shell – Groups 1 and 2.

The **p-block** contains those elements with outer electrons in the p sub-shell – Groups 4-8.

The **d-block** contains those elements with outer electrons in the d sub-shell – mostly the transition metals.

> **⬤** Z = number of electrons in an atom = number of protons

The **position** of the element in the Periodic Table immediately tells you how many electrons it has. This arises because the elements of each Group have the same outer electronic configuration. This is why elements of the same Group have similar chemical properties.

✓ *Quick check 1*

Worked example

Look at silicon, Si, in the Periodic Table. What is its electronic configuration?

Step 1: Silicon has full electron shells up to the configuration of neon, the noble gas in the Period above it.
We can represent this by writing [Ne] instead of $1s^2 2s^2 2p^6$.

Step 2: Silicon is in Group 4, so it has $ns^2 np^2$ as its outer electronic configuration.

Step 3: Silicon is in Period 3, so the electronic configuration of the outer electrons begins with 3.

Step 4: The full electronic configuration is therefore **[Ne] $3s^2 3p^2$**.

✓ *Quick check 2*

❓ *Quick check questions*

1 Explain why the p-block of the Periodic Table is given this name.

2 Write down the electron configurations of the elements A-D in the diagram opposite.

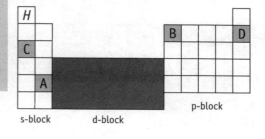

The Periodic Table of the Elements

Group

key

| relative atomic mass |
| atomic symbol |
| name |
| atomic number |

Example: 1.0 / H / Hydrogen / 1

1	2												3	4	5	6	7	0
																		4.0 He helium 2
6.9 Li lithium 3	9.0 Be beryllium 4												10.8 B boron 5	12.0 C carbon 6	14.0 N nitrogen 7	16.0 O oxygen 8	19.0 F fluorine 9	20.2 Ne neon 10
23.0 Na sodium 11	24.3 Mg magnesium 12												27.0 Al aluminium 13	28.1 Si silicon 14	31.0 P phosphorus 15	32.1 S sulphur 16	35.5 Cl chlorine 17	39.9 Ar argon 18
39.1 K potassium 19	40.1 Ca calcium 20	45.0 Sc scandium 21	47.9 Ti titanium 22	50.9 V vanadium 23	52.0 Cr chromium 24	54.9 Mn manganese 25	55.8 Fe iron 26	58.9 Co cobalt 27	58.7 Ni nickel 28	63.5 Cu copper 29	65.4 Zn zinc 30		69.7 Ga gallium 31	72.6 Ge germanium 32	74.9 As arsenic 33	79.0 Se selenium 34	79.9 Br bromine 35	83.8 Kr krypton 36
85.5 Rb rubidium 37	87.6 Sr strontium 38	88.9 Y yttrium 39	91.2 Zr zirconium 40	92.9 Nb niobium 41	95.9 Mo molybdenum 42	– Tc technetium 43	101 Ru ruthenium 44	103 Rh rhodium 45	106 Pd palladium 46	108 Ag silver 47	112 Cd cadmium 48		115 In indium 49	119 Sn tin 50	122 Sb antimony 51	128 Te tellurium 52	127 I iodine 53	131 Xe xenon 54
133 Cs caesium 55	137 Ba barium 56	139 La lanthanum* 57	178 Hf hafnium 72	181 Ta tantalum 73	184 W tungsten 74	186 Re rhenium 75	190 Os osmium 76	192 Ir iridium 77	195 Pt platinum 78	197 Au gold 79	201 Hg mercury 80		204 Tl thallium 81	207 Pb lead 82	209 Bi bismuth 83	– Po polonium 84	– At astatine 85	– Rn radon 86
– Fr francium 87	– Ra radium 88	– Ac actinium** 89	– Rf rutherfordium 104	– Db dubnium 105	– Sg seaborgium 106	– Bh bohrium 107	– Hs hassium 108	– Mt meitnerium 109	– Unn ununnilium 110	– Uuu unununium 111	– Uub ununbium 112			– Uuq ununquadium 114		– Uuh ununhexium 116		– Uuo ununoctium 118

lanthanides *

140 Ce cerium 58	141 Pr praseodymium 59	144 Nd neodymium 60	– Pm promethium 61	150 Sm samarium 62	152 Eu europium 63	157 Gd gadolinium 64	159 Tb terbium 65	163 Dy dysprosium 66	165 Ho holmium 67	167 Er erbium 68	169 Tm thulium 69	173 Yb ytterbium 70	175 Lu lutetium 71

actinides **

– Th thorium 90	– Pa protactinium 91	– U uranium 92	– Np neptunium 93	– Pu plutonium 94	– Am americium 95	– Cm curium 96	– Bk berkelium 97	– Cf californium 98	– Es einsteinium 99	– Fm fermium 100	– Md mendelevium 101	– No nobelium 102	– Lr lawrencium 103

Periodicity in the elements of Period 3

Element	sodium	magnesium	aluminium	silicon	phosphorus	sulphur	chlorine	argon
Symbol	Na	Mg	Al	Si	P	S	Cl	Ar
Physical state of the element	solid metal	solid metal	solid metal	solid metal	solid non-metal (P_4 molecules)	solid non-metal (S_8 molecules)	gas (Cl_2 molecules)	gas (Ar atoms)
Electronic configuration	$3s^1$	$3s^2$	$3s^2 3p^1$	$3s^2 3p^2$	$3s^2 3p^3$	$3s^2 3p^4$	$3s^2 3p^5$	$3s^2 3p^6$
Atomic radius/nm	0.191	0.160	0.130	0.118	0.110	0.102	0.099	0.095
Electrical conductivity	good	good	good	semi-conductor	poor	poor	poor	poor
Melting point/K	371	922	933	1683	317	386	172	84

This table shows us **trends** in the properties of the elements. A **trend** is a gradually changing event. The trends we see in this Period are repeated in all the Periods – this regular pattern of trends is called **periodicity**.

Now let's examine the trends across Period 3.

Atomic radius

Observation: The atomic radii of the elements decrease across the Period.

Explanation: The elements experience an increasing effective nuclear charge from left to right.

- This means that each electron added going across the period is in the same shell, so all the outer electrons of elements in the Period are approximately the same distance from the nucleus.

- The nucleus gains an extra proton for each element, so the nuclear charge is increasing and can act on each electron individually – the electrons are not shielded from the nuclear charge.

- This attraction pulls the electrons closer to the nucleus going across the Period so the atomic radius decreases.

Electrical conductivity

Observation: The trend is for decreasing electrical conductivity from left to right across the Period. The metals, on the left of the Period, are good conductors of electricity. The non-metals on the right are non-conductors. In between is silicon, which is a semi-conductor.

Explanation:

- Metals have metallic bonding which means positive ions with delocalised, mobile electrons around them. These mobile electrons will easily move if a potential is applied to the metal, so forming an electric current.

- P_4 and S_8 are solids, made up of molecules with covalent bonding linking the atoms in the molecules. However, the bonding between the molecules is weak. These substances cannot transfer electricity because the electrons are all located in the covalent bonds, where they are tightly held.

- Cl_2 is a gas – the molecules are too far apart to allow transfer of electrons.

- Silicon is a giant lattice of silicon atoms, and although the electrons are located in covalent bonds between the atoms, silicon can be made to transfer electrons under certain conditions – hence it is a semi-conductor.

✓ *Quick check 1*

Melting points

Observation: The melting points rise going across the Period to reach a maximum at silicon. Then there is a sharp drop to phosphorus, and then a general decrease to the end of the Period.

Explanation:

- The melting points of metals are high, reflecting the strong electrostatic bonding between positive ions and mobile electrons in the metallic bond. The melting points increase from Na→Mg→Al because the number of outer electrons increases from 1→2→3, so the electrostatic attraction is stronger and the metallic bond more difficult to break.

✓ *Quick check 2*

- The highest melting point is for silicon, which has a giant covalent structure. Covalent bonds are very strong, and require a lot of energy to be broken.

- The discrete molecules in phosphorus (P_4) and sulphur (S_8) have weak intermolecular bonds, which are easily broken and this means the melting points are much lower than for silicon, although they are all solid.

- Finally, as expected the gases Cl_2 and Ar have very low melting points.

Boiling points follow the same trend as melting points.

? Quick check questions

1 Sodium, aluminium and phosphorus are all elements found in Period 3 of the Periodic Table. Write down the order of electrical conductivity in these three elements. Explain why you have arranged them in this order.

2 Sodium, magnesium and aluminium are all metals but the melting point of sodium is much lower than the others. Explain why this is so.

Ionisation energy

Ionisation is all about removing electrons from atoms to make ions.

> The **first ionisation energy** is the energy required to remove 1 mole of electrons from 1 mole of gaseous atoms, to give 1 mole of gaseous ions.
>
> $$M(g) \rightarrow M^+(g) + e^-$$
>
> The **second ionisation energy** refers to the removal of the next mole of electrons from the mole of gaseous ions: $M^+(g) \rightarrow M^{2+}(g) + e^-$.
>
> The **third ionisation energy** refers to $M^{2+}(g) \rightarrow M^{3+}(g) + e^-$, and so on.

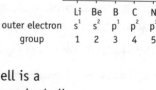 Learn this definition

Factors which influence the first ionisation energy

● **nuclear charge** – a large effective nuclear charge means the outer electron is difficult to remove, so the ionisation energy is large

● **atomic radius** – in larger atoms the outer electron does not feel such a great attraction for the nucleus, so it is easier to remove and the ionisation energy is low

● **electron shielding** – if more electrons shield the nuclear charge from the outer electron, it will be easier to remove, and the ionisation energy is low.

These facts mean that the ionisation energy

● **decreases down a Group** because the atomic radius and electron shielding both increase;

● **increases across a Period** because the effective nuclear charge increases.

✓ *Quick check 3*
✓ *Quick check 4*

Ionisation energies across a Period

Look at the graph of first ionisation energies of the elements in period 2:

There are 3 things to note about this graph:

● **There is a general increase across the period.** This is caused by the increasing effective nuclear charge.

● **There is a decrease between Groups 2 and 3.** This is because the Group 2 element has a full s^2 electronic configuration, whereas the Group 3 element has a $s^2 p^1$ electronic configuration. The p sub-shell is a higher energy level than the s sub-shell, so the single electron in the p sub-shell of elements in Group 3 is relatively easy to remove.

● **There is a decrease between Group 5 and Group 6.** Group 5 elements have a p^3 electronic configuration, whereas Group 6 elements have a p^4 electronic configuration. This fourth electron has to pair up with another electron, so repulsion comes into effect and this electron is relatively easier to remove.

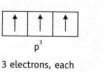

p^3

3 electrons, each
in a separate
p orbital

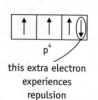

p^4

this extra electron
experiences
repulsion

How to predict electron configurations from ionisation energies

The values of successive ionisation energies can tell us the Group an element is from. **Successive ionisation energies** (IEs) simply means the 1st IE, the 2nd IE, the 3rd IE etc for one particular element.

Successive IEs for sodium look like this (see graph):

The 1st IE is relatively small, and then there is a large jump to the second IE. The rest are slightly larger still.

The electron configuration of sodium is $1s^2\ 2s^2\ 2p^6\ 3s^1$.

- The 1st IE refers to removing the $3s^1$ electron, and this is relatively easy to do because it is in the third principal shell, far away from the nucleus.

- The 2nd IE refers to removing the $2p^6$ electron, and this takes much more energy. This is because the electron is being removed from a full shell of electrons, which is a stable arrangement. Also, the electron is nearer the nucleus as it is in the second principal shell.

- The next two electrons to be removed are also in the second shell, and as the stability of the full shell has been destroyed they are not much more difficult to remove than the second electron.

Successive IEs for magnesium and aluminium look like this (see graphs):

Magnesium, in Group 2, has $3s^2$ and $3s^1$ electrons which are easy to remove, then a large increase in the amount of energy required to remove the next electron, $2p^6$, because it is in the lower principal shell.

Aluminium, in Group 3, has 3 electrons easy to remove in the outer shell, so 3 relatively low IEs.

We can see that the pattern of successive IEs for

- all Group 1 elements is 1 small, rest high
- all Group 2 elements is 2 small, rest high
- all Group 3 elements is 3 small, rest high

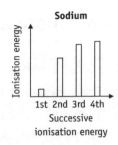

Sodium

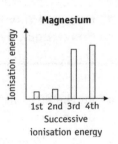

Magnesium

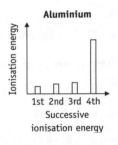

Aluminium

✓ *Quick check 1 & 2*

❓ *Quick check questions*

1 Sketch the shape of the bar chart of the first 6 successive ionisation energies you predict for the element carbon.

  Explain how this data shows that carbon is in Group 4.

2 Define **first ionisation energy**. Name three properties of the atom which influence the size of the ionisation energy. Explain why each property exerts an effect.

3 The first ionisation energy of arsenic, As, is 947 kJ mol^{-1}. Predict the value of the first ionisation energy of selenium, Se.

The Group 2 elements

Trends in the properties of the Group 2 elements magnesium, calcium, strontium and barium are studied in this section. The most important reactions of their compounds involve reduction and oxidation, which are defined in two new ways.

General properties

The Group 2 elements are all **reactive metals**.

- They are good conductors of heat and electricity.
- They are harder and denser than Group 1 metals, with higher melting points and stronger metallic bonding.
- Their **reactivity increases down the Group**.
- They are white in colour and relatively soft, and although a freshly-cut surface is shiny, it quickly oxidises in air and becomes dull.
- The electronic configurations of the elements are all the same - ns^2, where n is the Period number.   **Mg:** $2s^2$ **Ca:** $3s^2$ **Sr:** $4s^2$ **Ba:** $5s^2$
- All the elements have a full s sub-shell. This means they have similar physical properties.
- They all form ions with a 2^+ charge when they react.
- Their compounds are all white or colourless.
- They have similar chemical reactions.

Atomic radii

- The atomic radii **increase going down the Group**.
- This is because a new principal shell of electrons is present with each element, so the size of the atom increases.

✓ *Quick check 1*

Mg: 0.160 nm
Ca: 0.174 nm
Sr: 0.191 nm
Ba: 0.198 nm

Electrical conductivities

- All the Group 2 metals are good conductors of electricity because they have metallic bonding, which is composed of positive ions surrounded by mobile electrons.
- These electrons can move if a potential is applied because they are delocalised, so metals are good conductors.

Melting points (and boiling points)

Mg: 922 K
Ca: 1112 K
Sr: 1042 K
Ba: 998 K

- The melting points are all high, indicating strong metallic bonding.
- The bonding is strong because there are two electrons in the s sub-shell which become delocalised, so the positive ions have a charge of 2^+ and the electrostatic attraction is strong.
- The boiling points are very high too for the same reason.

The use of Group 2 elements and their compounds

Most of these involve magnesium and calcium.

Group 2 element or compound	Used in	Reason
magnesium	flares, tracer bullets, fireworks	magnesium burns with a bright white light
magnesium	rust prevention on ships	magnesium reacts in place of iron because it is more reactive
magnesium hydroxide $Mg(OH)_2$	antacids, indigestion remedies	it is a weak alkali so neutralises excess stomach acid
magnesium oxide	lining of furnaces	it has a very high melting point
magnesium fluoride	coating for camera lenses	reduces reflected light
calcium carbonate (limestone)	making cement, as marble	very hard and strong
calcium oxide (lime, quicklime)	mortar	very hard and strong
calcium oxide (lime, quicklime)	spread on acidic soil	it is basic so reduces the acidity of the soil and increases crop yields
calcium hydroxide – solid (slaked lime)	spread on acidic soil	as above
calcium hydroxide – aqueous (lime water)	laboratory test for CO_2 gas	it turns cloudy
calcium sulphate	plaster of Paris, used in plaster casts	it sets hard when mixed with water
barium sulphate	drunk by patients with gastrointestinal disease as a "barium meal"	it is opaque to X-rays so shows up the lining of the intestines when the patients is X-rayed

✓ Quick check 2

Compounds of calcium

There are some important compounds of calcium and you must know what they are, and some of their reactions.

Name	Calcium carbonate	Calcium oxide	Solid calcium hydroxide	Aqueous calcium hydroxide
Formula	$CaCO_3$	CaO	$Ca(OH)_2$ (s)	$Ca(OH)_2$ (aq)
Common name	limestone	lime	slaked lime	lime water
Reactions	decomposition: $CaCO_3(s) \rightarrow CaO(s) + CO_2(g)$	reaction with *water added dropwise* $CaO(s) + H_2O(l) \rightarrow Ca(OH)_2(s)$ A white lump of calcium oxide swells, steams and disintegrates as water is added, giving a white powder.	calcium hydroxide is sparingly soluble in water. The solution formed is mildly alkaline, pH 8–9, and is called lime water.	lime water has a pH of 8–9 and is used to test for the gas carbon dioxide. Carbon dioxide, when bubbled through lime water, produces a milky-white suspension – the lime water turns cloudy. $Ca(OH)_2(aq) + CO_2(g) \rightarrow CaCO_3(s) + H_2O(l)$

✓ Quick check 3

? Quick check questions

1 Suggest how the size of the Group 2 ions changes as the Group is descended.

2 State and explain why the main purchasers of commercially manufactured barium sulphate are Health Trusts.

3 When carbon dioxide gas is bubbled though aqueous calcium hydoxide a white suspension is seen. This reaction is used as a laboratory test for carbon dioxide gas. Construct an equation showing this reaction. State one other commercial use of calcium hydroxide.

Oxidation and reduction

Oxidation and reduction always occur at the same time. Reactions where this happens are called **redox reactions** (**red**uction and **ox**idation). You may have come across these definitions:

> **oxidation as the gain of oxygen; reduction as the loss of oxygen**
>
> **oxidation as the loss of hydrogen; reduction as the gain of hydrogen**

Now we will describe redox reactions in two new ways - **electron transfer** and **oxidation state (oxidation number).**

Electron transfer

Look at an oxidation reaction for a typical Group 2 element, magnesium:

$$2Mg \ (s) \ + \ O_2 \ (g) \ \rightarrow \ 2MgO \ (s)$$

In this reaction $Mg \rightarrow Mg^{2+} + 2e^-$, which shows us that magnesium has lost electrons. The magnesium has been **oxidised**. This means that

- **oxidation** is defined as **the loss of electrons**.

Similarly $O_2 + 4e^- \rightarrow 2O^{2-}$, which shows us that oxygen has gained electrons. The oxygen has been **reduced**. This shows that

- **reduction** is defined as **the gain of electrons**.

Remember **OILRIG:**
Oxidation Is Loss
Reduction Is Gain

✓ *Quick check 1*

Oxidation number (oxidation state)

Oxidation numbers (ON) are **the number of electrons that atoms lose, gain or share** when they form bonds. An oxidation number can be positive, negative or zero.

They are assigned to each atom in a compound or element. They are useful because

- **oxidation means an increase in oxidation number**
- **reduction means a decrease in oxidation number**

How to assign oxidation numbers

Follow these rules, and if two rules appear to contradict each other **follow the one that is higher in the list.**

- The oxidation number of an atom in the free element is zero.
- The total of the oxidation states of all the atoms in a molecule is zero.
- The oxidation number of monatomic ions is the same as the charge on the ion.
- In a polyatomic ion like SO_4^{2-}, then the total of the ONs of all the atoms in the ion equals the charge on the ion.
- In compounds, Group 1 metals have ON of = +1.
- In compounds, Group 2 metals have ON of = +2.

- In compounds, H has ON = +1.
- In compounds, F has ON = −1.
- In compounds, O has ON = −2.
- In compounds with metals, Cl, Br, I have ON = −1.
- In compounds with metals, Group 6 elements (O, S,) have ON −2.

✓ *Quick check 2*

Worked examples

What is the oxidation number of the underlined element in:

(a) <u>P</u>$_4$ This is the formula of the element phosphorus. Uncombined elements have ON = 0. We can write this as P$_4$ (0).

(b) <u>Al</u>$_2$O$_3$ The total of the ON of all the atoms = 0. The ON of oxygen, O = −2. The total for three O atoms is −6. The total for 2 Al atoms is +6.
Therefore ON of aluminium = **+3**.

(c) <u>Mn</u>O$_4^-$ The total of the ON of all the atoms is −1. The ON of oxygen, O is −2.
The total for four O atoms is −8.
The ON of manganese is **+7**.

(d) H$_2$<u>O</u> The total of the ON of all the atoms is 0. The ON of hydrogen = +1.
The ON of O is **−2**.

(e) Na<u>H</u> The total of the ON of all the atoms is 0. The ON of sodium is +1.
The ON of H is **−1**.

> ▶ Remember that oxidation numbers must always have a sign!

✓ *Quick check 3*

The **oxidation state** is another term for oxidation number. In systematic names the oxidation number is shown in brackets using roman numerals:
MnO$_4^-$ is manganate (VII)

? *Quick check questions*

1 State four ways of defining reduction. Give an example of one.

2 Work out the oxidation number of (a) sulphur in SO$_3^{2-}$, (b) nitrogen in NH$_4^+$, (c) carbon in CO$_3^{2-}$

3 Allocate oxidation numbers to each species in the following compounds:

NaCl; LiH; SrO; magnesium bromide; sodium hydroxide; silver nitrate; sodium sulphate, chlorine.

4 Use oxidation numbers to decide if the underlined species are oxidised or reduced in the reaction:

(a) <u>Mg</u>(s) + Cl$_2$(g) → MgCl$_2$(s)

(b) I$_2$ + 2S$_2$O$_3^{2-}$ → 2I$^-$ + S$_4$O$_6^{2-}$

(c) 2<u>Fe</u>$^{3+}$ + 2I$^-$ → 2Fe^{2+} + I$_2$

Reactions of the Group 2 metals

Reactions of the Group 2 metals with oxygen

All the metals burn – react with oxygen – to give the metal oxide. Once started, these reactions are vigorous! The oxide is a white solid.

Example: $2Mg(s) + O_2(g) \rightarrow 2MgO(s)$
ON   0   0   +2 –2

These are redox reactions, shown by looking at the oxidation number in the equation.

Magnesium is **oxidised:** it loses electrons, $Mg \rightarrow Mg^{2+} + 2e^-$ and the oxidation number increases from 0 to +2.

Oxygen is **reduced:** it gains electrons, $O_2 + 4e^- \rightarrow 2O^{2-}$ and the oxidation number decreases from 0 to –2.

Reactions of the Group 2 metals with water

All the metals react with water to give **hydrogen:**

Example: $Ca(s) + H_2O(l) \rightarrow Ca(OH)_2(s) + H_2(g)$
ON   0   0   +2 –2   0

✓ *Quick check 2*

The hydroxide salt formed is sparingly soluble, and can be seen as a cloudy white suspension.

Magnesium reacts very slowly with cold water – it can take a week to get a small test-tube full of hydrogen gas – but calcium reacts more quickly, and the others lower in the Group react more quickly still. This shows an important trend:

> **The Group 2 metals increase in reactivity as the Group is descended.**

- These are **redox reactions.** Calcium is **oxidised:** it loses electrons, $Ca \rightarrow Ca^{2+} + 2e^-$ and the oxidation number increases from 0 to +2.

Hydrogen is **reduced** as its oxidation number decreases from +1 in H_2O to 0 in H_2.

- Magnesium is very slow to react with cold water, but it reacts readily with steam:

$$Mg(s) + H_2O(g) \rightarrow MgO(s) + H_2(g)$$

Hydrogen gas is formed, like the reaction with water, but this time the oxide and not the hydroxide is formed. This is also a redox reaction, where magnesium is oxidised and hydrogen ion is reduced.

Explanation of the trend in reactivity

Magnesium is more reactive than calcium, which is more reactive than strontium etc. Why? This trend can be explained in terms of the **ionisation energies.** When they react, the Group 2 metals lose 2 electrons. This is more difficult in magnesium than calcium, because the outer electrons are **closer to the nucleus** with **less** shielding and so **held more tightly**. This is reflected in the ionisation energies, which are higher for magnesium than calcium.

Reactions of the Group 2 metals with hydrochloric acid

There are 3 reactions you must know:

> (1) magnesium with hydrochloric acid
>
> (2) magnesium oxide with hydrochloric acid
>
> (3) magnesium carbonate with hydrochloric acid

(1) magnesium + hydrochloric acid

The general reaction is

METAL + ACID → SALT + HYDROGEN

$$Mg(s) + 2HCl(aq) \rightarrow MgCl_2(aq) + H_2(g)$$

✓ *Quick check 1*

When a piece of magnesium is dropped into hydrochloric acid it immediately effervesces as hydrogen gas is given off. The magnesium disappears as it reacts and forms magnesium chloride, which is a colourless solution.

Oxidation numbers show that magnesium is oxidised and hydrogen is reduced – the chloride ion does not take part in the reaction.

✓ *Quick check 3*

(2) magnesium oxide + hydrochloric acid

The general reaction is

BASE + ACID → SALT + WATER

$$MgO(s) + 2HCl(aq) \rightarrow MgCl_2(aq) + H_2O(l)$$

This is not a redox reaction because the magnesium ion in MgO is not reduced to magnesium metal, and the oxidation numbers of all the species stay the same. No gas is given off in this reaction so there is no effervescence; the solid MgO reacts to give a colourless solution of aqueous magnesium chloride.

(3) magnesium carbonate + hydrochloric acid

The general reaction is

METAL CARBONATE + ACID → SALT + CARBON DIOXIDE + WATER

$$MgCO_3(s) + 2HCl(aq) \rightarrow MgCl_2(aq) + CO_2(g) + H_2O(l)$$

Again, this is not a redox reaction. Carbonates effervesce vigorously when added to acid as the carbon dioxide is released.

? *Quick check questions*

1 Explain, using oxidation states, whether magnesium is oxidised or reduced when it reacts with hydrochloric acid.

2 Explain what observations you would expect to make if a piece of calcium was dropped into warm water.

3 Calculate the volume of hydrogen produced at room temperature and pressure if 0.24 g of magnesium was reacted with an excess of sulphuric acid.

The Group 7 elements

The Group 7 elements you will study are chlorine, bromine and iodine. They all have a common electronic configuration of $ns^2 np^5$.

The Group 7 elements are also called the halogens, and they all occur naturally as diatomic molecules Cl_2, Br_2, I_2. They are non-metals. They are very reactive elements.

chlorine	$3s^2 3p^5$
bromine	$4s^2 4p^5$
iodine	$5s^2 5p^5$

Physical properties

- Chlorine is a gas, bromine is a liquid and iodine is a solid. This **decreasing trend in volatilities** of the halogens is caused by **increasing van der Waals' forces**.
- The halogens are **different colours**. Chlorine is a greenish-yellow gas, bromine is a dark red liquid which is volatile and gives off a dark red vapour, and iodine is a shiny grey-black crystalline solid. Iodine sublimes when heated gently – this means it goes straight from the solid to the vapour phase. Iodine vapour is purple.

✓ *Quick check 2*

Chemical properties

- Because the electronic configuration is $ns^2 np^5$ the halogens have to gain only one more electron to have a full p-orbital. This means that the most common oxidation number is zero for the elements and −1 for the ions. The ions are called **halide ions, Cl^-, Br^-, I^-**. Other oxidation states do exist, but they are not common; chlorine in particular can show a range of oxidation numbers, from −1 to +7.
- The trend in reactivity is that the halogens become **less reactive** on **descending** the Group.

When halogens react, they are reduced to the halide ion, so the substance they react with is oxidised. This means the halogens are **good oxidising agents**. Chlorine is the most powerful oxidising agent and iodine is the least powerful.

An oxidising agent oxidises another species and is itself reduced.

Relative reactivities

The relative reactivities (and oxidising powers) of chlorine, bromine and iodine can be shown by the reaction of the elements with (a) metals and (b) displacement reactions.

A reducing agent reduces another species and is itself oxidised.

(a) Reactions of the halogens with metals

When the halogens react with metals, they remove the outer electron from the metal to become a halide ion, and the metal becomes a positive ion.

Example: $2Na(s) + Cl_2(g) \rightarrow 2NaCl(s)$ $Na(0) \rightarrow Na(+1)$
ON   0   0   +1 −1 $Cl(0) \rightarrow Cl(-1)$

This is a **redox reaction**, where the halogen acts as an oxidising agent and oxidises the metal.

(b) Reactions with iron

The reactions of the halogens with **iron** illustrate the trend in reactivity nicely, because iron can be oxidised to Fe^{2+} or Fe^{3+}.

Chlorine is the strongest oxidising agent so can oxidise the iron as much as possible. Red-brown iron (III) chloride if formed.

$$2Fe(s) + 3Cl_2(g) \rightarrow 2FeCl_3(s) \qquad\qquad Fe(0) \rightarrow Fe(+3) \text{ oxidised}$$
$$Cl(0) \rightarrow Cl(-1) \text{ reduced}$$

Bromine is a weaker oxidising agent than chlorine so it does not oxidise all the iron to Fe^{3+}. Some of the iron is only oxidised to Fe^{2+}. A mixture of iron (III) bromide and iron(II) bromide is formed.

$$2Fe(s) + 3Br_2(g) \rightarrow 2FeBr_3(s) \qquad\qquad Fe(0) \rightarrow Fe(+3) \text{ oxidised}$$
$$Br(0) \rightarrow Br(-1) \text{ reduced}$$

$$2Fe(s) + Br_2(g) \rightarrow 2FeBr_2(s) \qquad\qquad Fe(0) \rightarrow Fe(+2) \text{ oxidised}$$
$$Br(0) \rightarrow Br(-1) \text{ reduced}$$

Iodine is a weak oxidising agent so does not produce any iron (III) chloride at all – only iron (II) chloride is formed. This reaction needs to be heated.

$$2Fe(s) + I_2(g) \rightarrow 2FeI_2(s) \qquad\qquad Fe(0) \rightarrow Fe(+2) \text{ oxidised}$$
$$I(0) \rightarrow I(-1) \text{ reduced}$$

> ▶ Remember –
> **Halogens** means the elements Cl_2, Br_2, I_2. **Halides** means the ions Cl^-, Br^-, I^-

Displacement reactions

In a displacement reaction a **more reactive element displaces a less reactive similar element.**

The full range of displacement reactions are:

Halide ion	Halogen		
	Cl_2	Br_2	I_2
Cl^-		no reaction	no reaction
Br^-	$Cl_2 + 2Br^- \rightarrow Br_2 + 2Cl^-$		no reaction
I^-	$Cl_2 + 2I^- \rightarrow I_2 + 2Cl^-$	$Br_2 + 2I^- \rightarrow I_2 + 2Br^-$	

> ✓ *Quick check 1*

These reactions are usually done in aqueous solution. However, it can be difficult to decide what is happening when you mix together aqueous chlorine and aqueous bromide ion, because the halide ions are all colourless in solution and the halogens have very faint colours (because they do not dissolve in polar water, as they are non-polar molecules). To see what is happening in these displacement reactions, we add **cyclohexane**. This is a non-polar organic solvent and it dissolves the halogens very well. This means **the halogens look coloured in cyclohexane** – chlorine is a faint green, bromine is brown-orange and iodine is purple.

> ▶ Cyclohexane is immiscible with water – it floats on top in a separate layer.

> **?** *Quick check questions*
>
> 1 State the observations you would make if you mixed bromine water with aqueous sodium iodide, and then added cyclohexane. Explain your observations by writing equations to illustrate any reactions which occur.
>
> 2 Explain, in terms on intermolecular bonding, why chlorine is a gas but iodine is a solid.

Reactions of the halogens

Chlorine is the most reactive halogen you study because it can capture an electron more easily than the others. Another way of saying this is:

- chlorine is the **strongest oxidising agent**

or

- chlorine is the **most electronegative** element

Chlorine is the most electronegative element (of Cl. Br, I) because it has the smallest atom. The outer p-orbital is closest to the nucleus so an electron from outside the atom can easily be transferred and held tightly in the p-orbital. This makes chlorine a strong oxidising agent.

> Solid metal halides are all white solids.

Laboratory test to determine which halide ion is in solution

The halide ions are all colourless in solution so a simple test is needed to find out which one is present in solution.

1 Acidify the unknown halide ion solution with nitric acid.

2 Add silver nitrate solution. A precipitate of the silver halide is formed:
- silver chloride is **white**
- silver bromide is **cream**
- silver iodide is **yellow**

3 It is difficult to accurately decide the colour of the precipitate, so ammonia solution is added:
- silver chloride dissolves in **dilute** ammonia solution - the others do not
- silver bromide dissolves in **concentrated** ammonia solution
- silver iodide does **not** dissolve even in concentrated ammonia solution

> ✓ *Quick check 3*

Two important reactions of chlorine

1 Reaction with cold, dilute aqueous sodium hydroxide

Chlorine reacts with dilute aqueous sodium hydroxide at room temperature to form **bleach**.

> The term "cold" is used in chemistry to mean *not heated – at room temperature.*

The formula of this bleach is NaClO, and it is called sodium chlorate (I). It is a salt which splits up in water to give free sodium ions and chlorate (I) ions, ClO^-. It is these chlorate (I) ions which have the bleaching properties – they oxidise stains and make them colourless.

$$Cl_2(g) + 2NaOH(aq) \rightarrow NaCl(aq) + NaClO(aq) + H_2O(l)$$

The ionic equation is

$$Cl_2 + 2OH^- \rightarrow Cl^- + ClO^- + H_2O$$

ON   0   -1   +1

Look at the oxidation numbers of chlorine in each of the three chlorine-containing species. The element chlorine has an oxidation number of 0, and when it reacts this goes up to +1 in ClO^-, and down to –1 in Cl^-. **The ON of one species is increased and decreased in the same reaction.** This is rather unusual. Reactions like this are called **disproportionation reactions**.

✓ Quick check 1

Most domestic bleach is a mixture of aqueous NaCl and NaClO. Stronger commercial bleach is usually HClO, not NaClO. HClO is called hypochlorous acid.

✓ Quick check 2

2 Reaction with water

Chlorine reacts with water to give **hydrochloric acid, HCl** and **hypochlorous acid, HClO.**

$$Cl_2(g) + H_2O(l) \rightarrow HCl(aq) + HClO(aq)$$

ON   0   -1   +1

This is also a **disproportionation reaction** – the oxidation number of chlorine increases and decreases in the same reaction.

This reaction is used to purify our water by removing bacteria and making it safe to drink. Water from the reservoir has solid particles removed from it, and then it is treated with chlorine. The reaction above shows us that acids are produced, and it is these that kill bacteria. The purified water is then piped to our homes. A small amount of chlorine remains in the water (some people can taste or smell it) to ensure it remains bacteria-free.

> ? **Quick check questions**
>
> 1 Chlorine reacts with concentrated sodium hydroxide solution at 70°C in this way:
>
> $$3Cl_2(g) + 6NaOH(aq) \rightarrow 5NaCl(aq) + NaClO_3(aq) + 3H_2O(l)$$
>
>   (a)   What is the oxidation state of chlorine in $NaClO_3$?
>
>   (b)   Explain why this is a disproportionation reaction.
>
> 2 Identify all the oxidation numbers in the ionic equation
> $$3ClO^- \rightarrow 2Cl^- + ClO_3^-.$$
> State and explain which species is oxidised and which is reduced.
>
> 3 Write instructions for a trainee analytical chemist who is identifying an unknown halide ion in solution for the first time.

Module A: End-of-module questions

1 This question looks at number relationships in Chemistry and the answer to each part is a number.

(a) State how many
- (i)   atoms there are in one molecule of butane
- (ii) neutrons there are in an atom of fluorine-19
- (iii) protons there are in an oxide ion
- (iv) electrons there are in the 3p sub-shell of a sulphur atom.   [4]

(b) Determine
- (i)   how many moles of ions there are in 2 moles of CaC
- (ii) the oxidation number of vanadium in the VO_3^- ion
- (iii) how many electrons there are in a NO_3^- ion.   [3]

(c) Calculate
- (i)   the relative atomic mass of naturally occurring gallium (^{69}Ga, 60%; ^{71}Ga 40%)
- (ii) how many grams of carbon there are in 8 g of CH_3OH
- (iii) how many grams of NaOH there are in 500 cm^3 of a 0.2 $mol.dm^{-3}$ solution.   [3]

2 Hydrogen fluoride, HF, is one of the most important of fluorine compounds. It can be prepared by reacting calcium fluoride, CaF_2, with sulphuric acid.

(a) (i)   Showing outer shell electrons only, draw 'dot-and-cross' diagrams of:
    1. Hydrogen fluoride   2. Calcium fluoride

(ii) Predict **two** differences between the physical properties of HF and CaF_2.   [4]

3. Wines often contain a small amount of sulphur dioxide that is added as a preservative. The amount of sulphur dioxide added needs to be carefully calculated; too little and the wine readily goes bad, too much and the wine tastes of sulphur dioxide. The sulphur dioxide content of a wine can be found by using its reaction with aqueous iodine.

$$SO_2(aq) + I_2(aq) + 2H_2O(l) \rightarrow SO_4^{2-}(aq) + 2I^-(aq) + 4H^+(aq)$$

(a) State the oxidation number of sulphur in SO_2 and in SO_4^{2-} [2]

(b) The sulphur dioxide content of a wine can be found by titration. An analyst found that the sulphur dioxide in 50 cm^3 of a sample of white wine reacted with exactly 16.4 cm^3 of 0.0100 $mol.dm^{-3}$ aqueous iodine. How many moles of
- (i)   iodine, I_2 did the analyst use in the titration?
- (ii) sulphur dioxide were in the 50 cm^3 of wine?
- (iii) What was the concentration, in mol dm^{-3} of sulphur dioxide in the wine?
- (iv) What was the concentration, in g dm^{-3} of sulphur dioxide in the wine?   [5]

Module B: Chains and Rings

This module deals with the basics of naming, formulae and isomerism and the chemistry of the homologous series of alkanes, alkenes, halogenoalkanes and alcohols. The chemistry of crude oil, fuels and addition polymerisation are dealt with separately.

The table below summarises the content and the previous knowledge and understanding required for efficient revision of the topics. If you cannot remember some of the facts from GCSE, do not worry: the important facts are covered here. For example, before the alkanes section it would be a good idea to look at the naming and isomers section first. It is likely that you would have partly covered this topic at GCSE and you will also need to refer to the Foundation module where intermolecular forces are covered.

Summary

Topic	Reference	Previous knowledge which you will use in this module
Formulae and isomerism	5.2.1	From GCSE – molecular formulae. You should already have done formulae from the Foundation module, but the terms, structural and displayed formulae are most used in Chains and Rings.
Naming	5.2.1	From GCSE – the names of the first four alkanes.
The alkanes	5.2.2	From GCSE – the names of the first four alkanes. From Foundation – sigma (σ) bonding.
Hydrocarbons and fuels	5.2.3	This follows on from the alkanes. From GCSE – fractional distillation and cracking, There is, however, a lot of new stuff here.
The alkenes	5.2.4	From GCSE– Some facts could be recalled about unsaturated hydrocarbons and displayed formulae. From Foundation - sigma (σ)and pi (π) bonding and the shapes of molecules.
Addition polymerisation	5.2.4 and 5.4.6	Some of this may have been covered at GCSE and there is not a lot of new stuff really but you should have a good understanding of the alkenes section.
The alcohols	5.2.5	From GCSE – you may recall the fermentation process. From Foundation – hydrogen bonding.
Infrared spectroscopy	5.2.5	The section on the alcohols and the work on the oxidation of the alcohols to aldehydes, ketones and carboxylic acids.
The halogenoalkanes	5.2.6	From GCSE – the identity of the halogens. The naming of these compounds should be understood.
Yields	5.2.1	From Foundation and GCSE – moles and reacting masses.
Questions		Review the module.

Formulae and isomerism

This important section deals with two basics of organic chemistry – **types of formulae** and the **isomerism** of organic compounds. You should know about empirical and molecular formulae as covered in the Foundation section. **Skeletal formulae** and **positional isomerism** will be dealt with in more detail in the A2 book, in the section on arenes.

Types of formulae

- **The empirical formula** is the simplest whole number ratio of elements in a compound. For example, all alkenes have the empirical formula CH_2. The molecular formulae for ethene (C_2H_4) and propene (C_3H_6) can be simplified to CH_2.

- **Molecular formulae** give the actual number of each kind of atom present. For example, $C_4H_{10}O$ is the molecular formula of the alcohol butanol with 4 carbon atoms, 10 hydrogen atoms and 1 oxygen atom.

- **General formulae** represent any member of a homologous series. For example, the alkane with n carbons has the formula C_nH_{2n+2}.

- **Structural formulae** are better representations of the molecule because they give an idea of how the carbon atoms and other groups are arranged. For example butanol, $C_4H_{10}O$, can exist in three different forms, all with the same molecular formula but with different arrangements of the atoms,

$CH_3CH_2 CH_2CH_2OH$	$CH_3CH_2CH(OH)CH_3$	$CH_3C(CH_3)(OH)CH_3$
butan-1-ol	butan-2-ol	2-methyl propan-2-ol

- **Displayed formulae** give the most accurate representation of the molecule in that they show all the bonds present and can give some indication of the different ways the atoms are arranged in space. For example, propene can be represented as shown in the margin rather than $CH_3CH=CH_2$.

propene

✓ *Quick check 1, 3, & 4*

- **Skeletal formulae** are most often used when representing cyclic compounds such as cycloalkanes and arenes.

Isomerism

> **Isomers of a compound have the same molecular formulae but have different structural and displayed formulae.**

▶ Learn this definition

There are TWO different types of isomerism.

- In **STRUCTURAL ISOMERISM** the structural formulae are different. The **carbon chain** may be different or the **functional group** and/or its **position in the molecule** may be different.

✓ *Quick check 2*

- In **STEREOISOMERISM** the structural formulae are identical but the displayed formulae are different because the isomers have different arrangements of the atoms in space. There are two types, *cis-trans* isomers (see alkenes section) and optical isomers (this is part of the A2 specification and is not required for AS level).

Structural isomerism

There are two main types you will need to know about. Remember the molecular formula is the same for any set of isomeric compounds.

✓ Quick check 4

- **Chain isomerism.** Here the functional group is the same but the arrangement of the carbon atoms in the chain is different.
- **Functional group isomerism.** The functional group is different.

✓ Quick check 5

- **Positional isomerism.** The position of the functional group differs. This is treated in more detail in the arenes section in the A2 specification.

Examples of chain isomerism

✓ Quick check 1,3&4

Molecular formula	Isomers		Type of isomerism
1. C_4H_{10}	$CH_3CH_2CH_2CH_3$	butane	Chain
	$CH_3CH(CH_3)CH_3$	2-methylbutane	
2. C_3H_7Cl	$CH_3CH_2CH_2Cl$	1-chloropropane	Positional
	$CH_3CHClCH_3$	2-chloropropane	
3. C_4H_9OH	$CH_3CH_2CH_2CH_2OH$	butan-1-ol	Chain and positional
	$CH_3CH_2CH(OH)CH_3$	butan-2-ol	
	$CH_3C(CH_3)(OH)CH_3$	2-methylpropan-2-ol	
	$CH_3CH(CH_3)CH_2OH$	2-methylpropan-1-ol	
4. C_3H_6O	CH_3CH_2CHO	propanal (an aldehyde)	Functional group isomerism
	CH_3COCH_3	propan-2-one (a ketone)	

❓ Quick check questions

1 Draw the displayed formulae and name the isomers of the following compounds.

   (a) C_5H_{12}

   (b) C_3H_7OH

2 For the isomers of C_3H_7OH explain what type of isomerism is shown.

3 Draw as many displayed isomers as you can of the following compounds.

   (a) $C_2H_4Cl_2$

   (b) C_6H_{14}

4 C_5H_{10} has 5 isomers. Two of them are *cis-trans* isomers (see alkenes section). Draw the displayed formulae for all five isomers and name them.

5 The two isomers of C_2H_6O are ethanol (C_2H_5OH) and dimethyl ether (CH_3OCH_3). Explain what kind of isomerism shown by these two compounds.

Naming organic compounds

Naming organic compounds using their formulae or constructing their structural and displayed formulae from their names is an important part of organic chemistry and the 'Chains and Rings' syllabus. Names are based on the I.U.P.A.C. convention and we will look at the naming of all the groups of compounds found on your AS specification.

Naming alkanes

- Their names always end in the suffix **–ane**.
- The first four alkanes are methane, ethane, propane and butane.
- Substituent groups are named according to the alkane from which they are derived. For example the CH_3- group is derived from methane and is called the **methyl** group. Similarly C_2H_5- comes from ethane and is called the **ethyl** group. These group names are placed at the beginning of the name.

How is it done?

Worked example

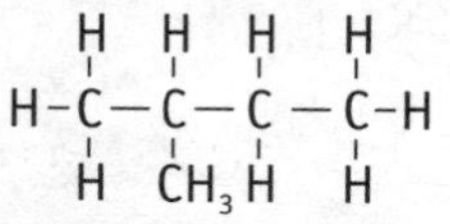

```
        H   H   H   H
        |   |   |   |
    H – C — C — C — C–H
        |   |   |   |
        H  CH₃ H   H
```

The steps in naming this alkane are as follows:

Step 1: Identify the longest carbon chain. In this case it is four carbons. Therefore the parent compound is butane.

Step 2: Number the carbon (the lowest possible) to which the substituent group is attached. Here it is the second carbon and therefore the substituent group is indicated by the prefix, 2-methyl. If there are more than one substituent groups then the numbers of the carbon atoms to which they are attached are indicated in the formula.

Step 3: Put the two bits together and name the whole compound. In this case, 2-methylbutane.

Examples

Structural formula	Longest carbon chain	Identity and position of substituent group	Name
$CH_3CH(CH_3)CH_3$	THREE	Methyl group on second carbon	2-methylpropane
$CH_3C(CH_3)_2CH_3$	THREE	2 methyl groups on second carbon	2,2,dimethyl propane
$CH_3CH(CH_3)CH_2CH_3$	FOUR	Methyl group on second carbon	2-methylbutane

Naming halogenoalkanes

Step 1: Find the longest carbon chain

Step 2: Put the position (lowest number possible) and name of the halogen in front. The halogen is always halo, e.g. bromine is bromo-

✓ *Quick check 1,2&3*

Examples

Structural formula	Longest carbon chain	Identity and Position of halogen atom	Name
$CH_3CH_2CHICH_3$	FOUR	Iodine on second	2-iodobutane
$CH_3CH_2CH_2Br$	THREE	Bromine on first	1-bromopropane
$CH_3CH(CH_3)CH_2Cl$	THREE	Chlorine on first	1-chloro-2-methylpropane

Naming alkenes

Step 1: Find the longest carbon chain and change the ending from ane to ene.

Step 2: The lowest possible position of where the C=C bond starts is indicated by the number in the name.

✔ *Quick check 1&3*

Examples

Structural formula	Longest carbon chain	Position of double bond	Name
$CH_3CH=CH_2$	THREE	First carbon	Propene
$CH_3CH_2CH=CH_2$	FOUR	First carbon	But-1-ene
$CH_3CH_2CH=CHCH_3$	FIVE	Second carbon	Pent-2-ene

Naming alcohols

Step 1: Find the longest carbon chain and name the alkane (without the –e).

Step 2: The lowest position of the –OH group is indicated by a number followed by –ol.

✔ *Quick check 1,2&4*

Examples

Structural formula	Longest carbon chain	Position of –OH group	Name
$CH_3CH(OH)CH_2$	THREE	Second carbon	Propan-2-ol
$CH_3CH_2CH_2CH_2OH$	FOUR	First carbon	Butan-1-ol
$CH_3CH(CH_3)CH_2OH$	FIVE	First carbon	2-methylpropan-1-ol

❓ *Quick check questions*

1 Name the following compounds

   (a) $CH_3CH_2CH(OH)CH_3$ (b) $CH_3CH_2CH_2Cl$

   (c) $CH_3CH=CHCH_3$ (d) $CH_3CHBrCH_3$

2 Draw the displayed formulae of the following compounds:

   (a)   2-chloro-butane   (b)   propan-1-ol

   (c)   2-methylpropan-2-ol   (d)   2-chloropropene

3 Why is it that for bromoethane and propene no numbers are required to indicate the positions of the halogen atom or double bond?

4 Draw the **four** structural isomers of the alcohol C_4H_9OH and name them.

The alkanes

Background facts

- They are a **homologous** series of **hydrocarbons** and therefore the members of the series have the same general formula (C_nH_{2n+2}). For example, in ethane $n=2$ and therefore there are $2\times2+2 = 6$ hydrogens and its molecular formula is C_2H_6.

  As with all homologous series the formulae of successive members of the series differ by $-CH_2$.

- They have **no multiple bonds in their molecules** and are therefore **saturated hydrocarbons**.

- All the bond angles are 109°28′ (109·5°) and each carbon atom is the centre of a tetrahedron with other atoms at each of the four corners.

- The only bonds present in their molecules are either C-C or C-H bonds. Neither of these bonds are polar making them unreactive molecules which have few reactions.

- Because they are non-polar molecules the forces between their molecules are weak **van der Waals' forces**.

- Their main uses are as fuels and as feedstock for the production of other chemicals (see (a) to (c) below).

  For example, methane is the main constituent of natural gas and is burned to produce heat. It is also used in the Haber process to make hydrogen for the production of ammonia.

  Longer chain alkanes are **cracked**, to give alkenes and shorter alkanes that are used as fuels.

✓ *Quick check 1*

> These forces increase with increasing size of alkane.

Combustion

- All alkanes undergo complete combustion (burn in excess air containing oxygen) to give carbon dioxide and water.

  e.g. $CH_4(g) + 2O_2(g) \rightarrow CO_2(g) + 2H_2O(l)$

✓ *Quick check 2*

- If there is poor ventilation then the lack of oxygen causes incomplete combustion and poisonous carbon monoxide is formed along with water.

  e.g. $CH_4(g) + 3/2O_2(g) \rightarrow CO(g) + 2H_2O(l)$

- After combustion in the car engine there are several pollutants in the car exhaust fumes. These are removed using a **catalytic converter** (see page 75). This is a platinum–rhodium catalyst on a high surface area ceramic honeycomb. The catalyst changes the pollutants to harmless products as shown in the table:

Pollutant	How is it formed?	What is it converted to?
Hydrocarbons	It is unburned fuel.	CO_2 and H_2O
Carbon monoxide	Incomplete combustion of hydrocarbons	CO_2
Oxides of nitrogen (NO_x)	Reaction between oxygen and nitrogen in the air in the car engine.	Oxygen and nitrogen.

The melting and boiling points of the alkane

✓ *Quick check 3*

- The longer the carbon chain, the more points of contact and the stronger the van der Waals' forces.

- When the alkanes are branched there are fewer points of contact and consequently the van der Waals' forces are weaker.

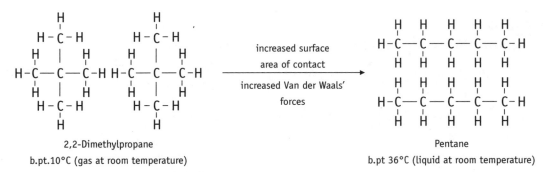

2,2-Dimethylpropane
b.pt.10°C (gas at room temperature)

increased surface
area of contact

increased Van der Waals'
forces

Pentane
b.pt 36°C (liquid at room temperature)

Free-radical substitution

In this reaction, a hydrogen on the alkane is replaced by (substituted by) chlorine or bromine. The overall reaction for methane is

$$CH_4(g) + Cl_2(g) \rightarrow CH_3Cl(l) + HCl(g)$$

The product is always a **halogenoalkane**.

There are three stages in this reaction mechanism

Initiation $Cl_2 \rightarrow 2Cl\bullet$ ($\bullet$ on the chlorine indicates a free radical)

The free radicals are produced at this stage. It always involves the cleavage of the Cl–Cl covalent bond by UV light. Because two free radicals are formed by the splitting, this is called **homolytic fission.**

Propagation   This stage keeps the reaction going because it produces free radicals making it into a chain reaction. Examples of reactions at this stage are:

$CH_4 + Cl\bullet \rightarrow CH_3\bullet + HCl$ ($CH_3\bullet$ is a methyl radical)

$CH_3\bullet + Cl_2 \rightarrow CH_3Cl + Cl\bullet$

Termination   This ends the chain reaction because free radicals are 'mopped up'.

$CH_3\bullet + CH_3\bullet \rightarrow C_2H_6$ (ethane)

$CH_3\bullet + Cl\bullet \rightarrow CH_3Cl$ (chloromethane)

> The word **homolytic** means to break down (lysis) into the same (homo). **Fission** means to split.

✓ *Quick check 4*

? Quick check questions

1 Give the molecular formula for the alkane, octane.

2 Write a balanced symbol equation for the combustion of pentane.

3 Butane can form two isomers.

   (a) Draw their displayed formulae and name them.

   (b) (i) Which isomer has the higher boiling point.

       (ii) Explain your answer.

4 One of the products of the free radical substitution of methane by chlorine is C_2H_6. Explain how you think this might have been formed.

Hydrocarbons and fuels

Crude oil is a source of important chemicals and fuels and several processes are used to carry out the changes.

What is a good fuel?

Petrol (a mixture of readily available, low boiling point liquids) is an example of a good fuel. There are several reasons why:

- It's vapour mixes easily with air/oxygen. The temperature at which its vapour can ignite – its flash point, is very low.
- Because it is liquid, it is easily transported and easy to store.
- It leaves little or no residue after it has burned.
- When it burns it releases large amounts of energy and the gases formed have a much greater volume than the reactants. This large volume change can be used to push a piston in a car engine.
- The structures of the liquids in the petrol mixture can be changed so that they burn steadily and smoothly.

✓ *Quick check 1*

Facts about crude oil and other fossil fuels

- The three fossil fuels are coal, oil and natural gas.
- They were formed over millions of years from the remains of dead animals and plants. They cannot be replaced and are therefore **non-renewable**.
- They are also valuable sources of other important chemicals which are either useful themselves or as starting materials (**feedstock**) in other chemical processes. Methane for example, is used in the Haber process for making hydrogen.
- Crude oil is mostly a mixture of **alkanes** with different boiling points.

The processes used in the petroleum industry

1 Fractional distillation

- This separates the components of the crude oil on a fractionating tower on the basis of their different boiling points.
- The **heavier fractions** with high boiling points condense first and therefore come off lower down the column; **lighter fractions** come off higher up.

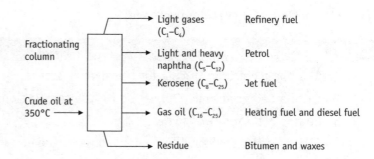

Fractionating column		
	Light gases (C_1–C_4)	Refinery fuel
	Light and heavy naphtha (C_5–C_{12})	Petrol
	Kerosene (C_8–C_{25})	Jet fuel
Crude oil at 350°C	Gas oil (C_{16}–C_{25})	Heating fuel and diesel fuel
	Residue	Bitumen and waxes

2 Cracking

- We get only 20% of the petrol required for car fuel from fractional distillation. Therefore another process is needed to make up the shortfall.

✓ *Quick check 3*

- Cracking uses heat and a catalyst to break up longer chain alkanes molecules into alkenes (used in polymerisation) and either more useful shorter alkanes (used as car fuels) or hydrogen. It is a type of thermal decomposition.

Examples: $C_{10}H_{22}(l) \rightarrow C_8H_{18}(l) + C_2H_4(g)$

$C_4H_{10}(g) \rightarrow C_3H_6(g) + H_2(g)$

> Equations for cracking are easily balanced if you know the products. Alkanes have the general formula, C_nH_{2n+2} and alkenes have the general formula, C_nH_{2n}.

✓ *Quick check 2&3*

3 Isomerisation

- In a car engine, straight chain alkanes ignite prematurely, causing 'knocking', which could damage the engine. They have low octane numbers.

- If the molecule is branched less knocking occurs.

- Isomerisation uses a catalyst to convert straight chain alkanes into branched chain alkanes.

Example: $CH_3(CH_2)_6CH_3 \rightarrow CH_3CH(CH_3)CH_2C(CH_3)_3$

Octane   2,2,4-trimethylpentane

> The higher the octane number, the less the amount of knocking. The octane numbers of zero for heptane to 100 for 2,2,4-trimethylpentane are used as standards. N-octane has a value of −20 whilst benzene's is 120.

✓ *Quick check 2&3*

4 Reforming

- This is another way to increase the octane rating.

- It needs a catalyst and a moderately high temperature.

- The straight chain alkane loses hydrogen to form a cyclic alkane and then an arene.

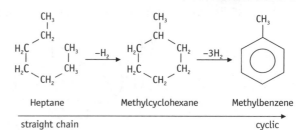

Heptane   Methylcyclohexane   Methylbenzene

straight chain   cyclic

Alternative fuels

Biogas is fuel made from the anaerobic digestion of waste material such as animal manure and dead plants. Its main constituent is methane.

> The structures for cyclohexane and benzene can be represented using skeletal formulae:

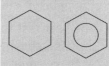

? Quick check questions

1 Write the equation for the combustion of heptane (C_7H_{16}) and explain why the reaction causes the volume to increase.

2 Octane from crude oil can be converted into 2,2,4-trimethylpentane and 1,4-dimethylbenzene ($C_6H_4(CH_3)_2$). Name the processes used to carry out these conversions and describe what happens in each one.

3 (a) Write equations for the cracking of dodecane ($C_{12}H_{26}$) into

    (i)   Decane and ethene.

    (ii)   Octane and butene.

  (b) The decane and octane are straight chain compounds.

    (i)   Why are these compounds not used in this form as fuels in the car engine?

    (ii)   What process is now carried out in order to make them useable as car fuels?

The alkenes

The alkenes are an extremely important homologous series. Their reactions give you a first good insight into reaction mechanisms and they are the building blocks for the manufacture of polymers and hence plastics.

Background facts

- They are a **homologous series** of **hydrocarbons** and the **functional group** is the **C=C** in their molecules. Each member of the series differs from the next one by CH_2-

$$CH_2 \text{ difference} \left(\begin{array}{ll} CH_2=CH_2 & \text{ethene} \\ CH_3CH=CH_2 & \text{propene} \end{array} \right.$$
$$CH_2 \text{ difference} \left(\begin{array}{ll} CH_3CH_2CH=CH_2 & \text{butene} \end{array} \right.$$

- Because the C=C bond is a multiple bond it makes them **unsaturated** compounds. Their reactions are characteristic of this **functional group**.

- Their general formula is C_nH_{2n}. For example, ethene has 2 carbons and therefore 4 hydrogens.

✓ *Quick check 1*

- The C=C bond consists of a σ (sigma) bond and a π (pi) bond (see below).

- They undergo **addition** reactions where the alkene reacts with another compound to form one single product molecule.

  e.g. $CH_2=CH_2(g) + Br_2 (l) \rightarrow CH_2Br\text{-}CH_2Br(l)$

      ethene   bromine   1,2 dibromoethane

- In addition reactions they react with electron pair acceptors called **electrophiles**. Examples of electrophiles are polarised Br_2 and H^+ in HBr (or HCl).

- The best test for an alkene is the reaction with bromine water in the dark. The bromine changes from orange to colourless as an addition product is formed. For example, propene reacts as follows to form 1,2 dibromopropane – a dihalogenoalkane:

  $CH_3CH=CH_2(g) + Br_2(aq) \rightarrow CH_3CHBr\text{-}CH_2Br$

                      Orange   Colourless addition product

- They exhibit *cis-trans* (**geometric**) isomerism, which is a type of stereoisomerism. For example but-2-ene has 2 isomers – a *trans* and a *cis* isomer.

- Their main use is in the production of **addition polymers.** For example ethene forms polyethene, propene forms polypropene (see polymer section for more details on this).

> Alkanes will decolorise bromine in the presence of light so to be precise we have to stipulate that the reaction takes place in the dark.

Shape and bonding

- The σ bond lies along the line (axis) between the C=C carbons.
- The π bond is formed by the sideways overlap of 2 spare p-orbitals.
- The π bond lies above and below the plane of the flat alkene molecule.
- The π electrons do not contribute to the shape. Round each carbon atom there are three pairs of σ-electrons and therefore the shape around each one is **trigonal (triangular) planar**.
- The bond angles are 120°.

✓ *Quick check 2*

Cis-trans isomerism

- In alkenes there is no free rotation about the C=C bond and this can produce *cis-trans* isomerism (geometric isomerism).
- If there are two different groups on each carbon of the C=C bond then we can get stereoisomers.
- These have the same structural formula, but are arranged differently in space with respect to the C=C bond.
- The two isomers are *cis-* and *trans*-isomers.
- The trans-isomer has similar groups diagonally opposite to each other across the C=C bond. The *cis-* isomer has them on the same side of the C=C bond.

✓ *Quick check 3*

Example: But-2-ene.

Structural formula $CH_3CH=CHCH_3$.

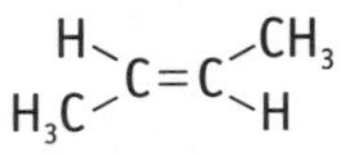

groups on opposite sides   groups on same side
trans-but-2-ene   *cis*-but-2-ene

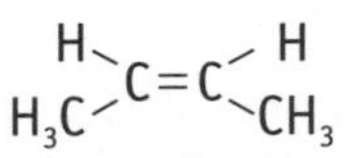

Quick check questions

1 Draw the displayed formulae for the four isomers of butene (C_4H_8).
2 Explain why the bond angles in ethene and other alkenes are 120°.
3 (a) What type of isomerism is exhibited by alkenes and not by alkanes?
   (b) Explain why this is a form of stereoisomerism and not structural isomerism.
   (c) Which of the following compounds exhibit this form of stereoisomerism? Explain your answers.
        (i) $(CH_3)_2C=CH_2$ (ii) $CH_3CH_2CH=CHCH_3$

Reactions of alkenes

- The π bond produces two regions of high electron density in the molecule. In reactions this attracts positive ions or even induces dipoles in some molecules (e.g. Br_2)
- When alkenes react, the electrons in the π bond form a covalent bond with the positive part of the attacking molecule. The mechanism for bromine is shown below.
- Because the alkene **adds** to an **electrophile,** the mechanism is called an **electrophilic addition.**

In this reaction the electron-deficient bromine accepts a pair of electrons and therefore acts as an **electrophile.**

✓ Quick check 3

The mechanism below shows the sequence of reactions for an alkene and bromine. The 'curly arrows' mean that a pair of electrons is 'on the move'.

UNSATURATED
SATURATED
δ+
δ−
Br − Br
Br − Br
C − Br
C⁺
Br⁻
C − Br
C − Br

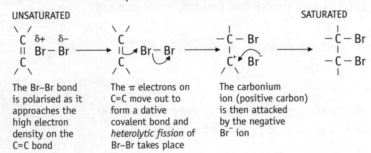

The Br–Br bond is polarised as it approaches the high electron density on the C=C bond

The π electrons on C=C move out to form a dative covalent bond and *heterolytic fission* of Br–Br takes place

The carbonium ion (positive carbon) is then attacked by the negative Br^- ion

- You have to think of these addition reactions as reactions with X-Y as it adds across the double bond.
- X bonds with one carbon whilst Y bonds with the other.
- Because X-Y splits to form X^+ and Y^- this is called **heterolytic fission** (splitting)

Examples are:

Molecule	X	Y
Br_2	Br	Br
H_2	H	H
H_2O	H	OH
HBr (or HCl)	H	Br (or Cl)

General example

✓ Quick check 1,2 & 4

Addition
C=C
X−Y
Unsaturated
C−C
X Y
Saturated

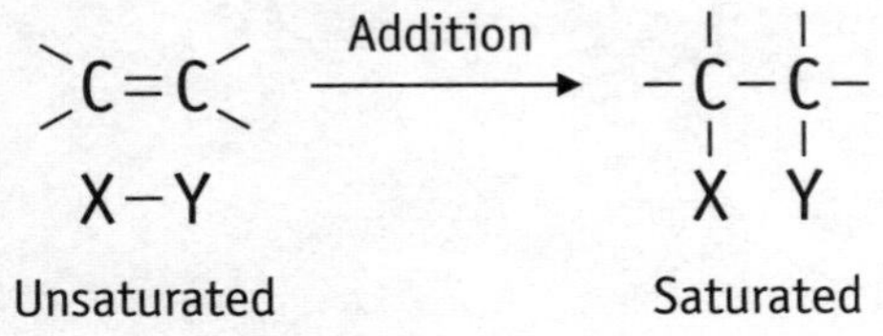

A summary of the reactions of propene

The reaction with hydrogen is used in the food industry to convert polyunsaturated vegetable oils to margarine.

✓ *Quick check 4*

This is the best test for alkenes. The bromine changes from orange to colourless.

NOTE for asymmetric alkenes like propene, there are two isomeric products when X and Y are different, as in the addition of HOH (H_2O) and HBr.

Exam tip: When drawing polymers, you are often asked to draw at least 4 carbons with dashed lines at each end to show that the chains carry on.

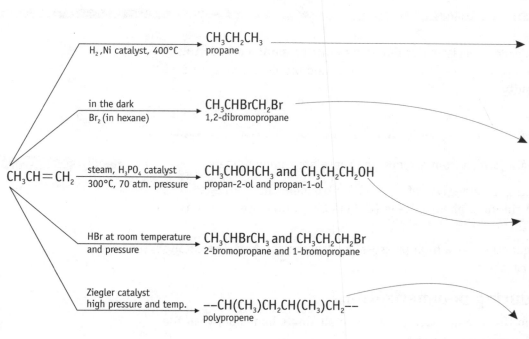

? *Quick check questions*

1 Give the conditions and the names and structural formulae of the products of the reactions of

(a) ethene and (b) *trans*-but-2-ene with the following:

(i) water (steam)

(ii) hydrogen

(iii) hydrogen bromide

(iv) bromine?

2 A hydrocarbon X consisted of 85.71% carbon and 14.29% hydrogen. Its relative molecular mass was shown to be 42 $g.mol^{-1}$.

(a) Calculate its empirical formula

(b) Calculate its molecular formula and draw its displayed formula.

(c) Draw the two products from X's reaction with steam.

(d) Draw the two products of X's reaction with hydrogen bromide.

3 In the reaction mechanism for the addition of bromine to ethene, one of the bromine atoms is electron deficient and acts as an ***electrophile***. Explain the word in *italics*.

4 Animal fats are mainly saturated fats. Explain why they cannot be used for the production of margarine.

Addition polymerisation

Polymers are large molecules consisting of repeating units of **monomers**. In this section you will revise how they are formed, their uses and problems associated with their non-biodegradability.

Background facts

✓ Quick check 1

- Addition polymers are formed from alkenes and these are the monomers.
- Many monomer units are derivatives of ethene. For example chloroethene ($CHCl=CH_2$) or vinyl chloride, phenylethene ($C_6H_5CH=CH_2$) otherwise known as styrene.
- Polymerization frequently uses a high pressure, heat and a catalyst. The catalyst is called a **Ziegler catalyst**.

What happens during polymerization?

- The double bonds in the alkenes open out to form all single bonds between the monomers and they are linked together.
- The electrons in the π-bonds form the new σ-bonds between the monomer units.
- How this happens for ethene is shown in the diagram below.

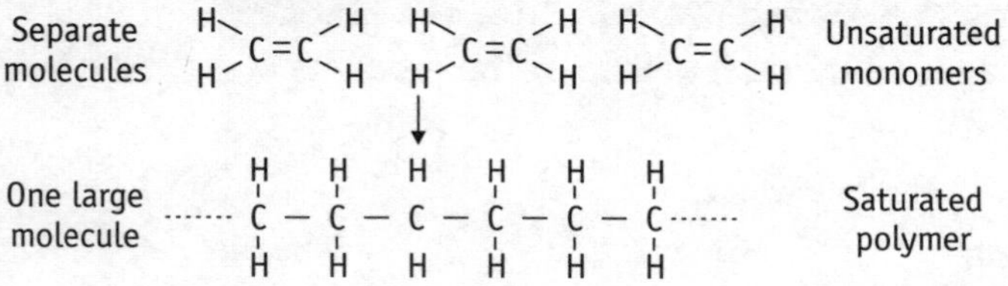

The easiest way to work out the structure is to think of the monomer as the repeat unit shown below. When you draw the structure of the addition polymer you either draw at least FOUR carbons as shown below, or indicate a repeating structure as –[monomer unit]$_n$–

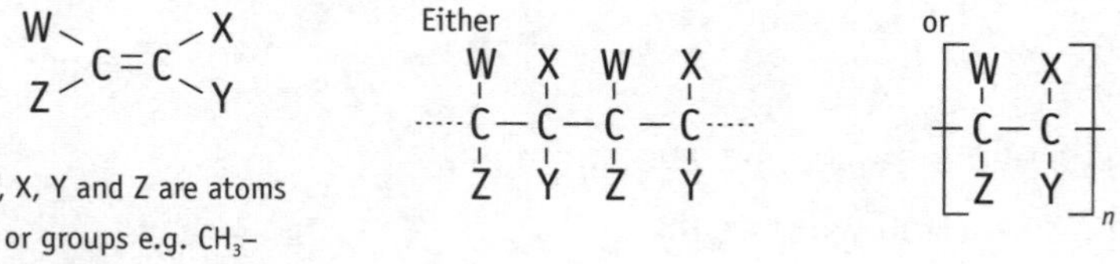

Examples

Formula of monomer	Name of monomer	Polymer formed
$CH_3CH=CH_2$	Propene	Polypropene —$CH(CH_3)$-CH_2-$CH(CH_3)$-CH_2— or –[$CH(CH_3)CH_2$]$_n$-
$CHCl=CH_2$	Chloroethene(vinyl chloride)	Polychloroethene (PVC) —$CHCl$-CH_2CHCl-CH_2— -[$CHClCH_2$]$_n$-
$CF_2=CF_2$	Tetrafluoroethene	Polytetrafluoroethene (PTFE) —CF_2-CF_2-CF_2-CF_2— -[CF_2-CF_2]$_n$-

Problems with polymers

- Polymers like polyethene contain either non-polar (e.g. C–H bond) or strong bonds (e.g. C–F bond), making them unreactive and because of this they are **non-biodegradable**. Therefore they persist in the environment and cause a litter problem.

✓ Quick check 3

- The best way to deal with polymers is to recycle them but this can be difficult because the different types are difficult to separate. Plastics made from recycled polymers are of a lower quality.

- If we try to dispose of them by burning they give toxic products like carbon monoxide or hydrochloric acid depending on the polymer's structure. However, in efficient furnaces they are good fuels. It is particularly hazardous to burn polyvinylchloride because toxic compounds like HCl and PCBs (polychlorinated-biphenyls) are formed.

✓ Quick check 2 & 3

- New processes are being developed by which polymers can be cracked to give new alkenes. Also, in Japan, recycled plastics are being used as fuels in blast furnaces and many power generators use them in a similar way.

? Quick check questions

1 (a) Draw the displayed formulae of the following monomers.

   (i)   ethene

   (ii)   propene

   (iii) chloroethene

   (iv)   tetrafluoroethene

   (b) For each monomer draw the resulting polymer, showing at least four carbons.

2 For the each of the following polymers give the possible polluting products of combustion.

   (a) polyethene

   (b) polychloroethene

3 (a) Draw the monomer unit, which would lead to the formation of the polymer shown below.

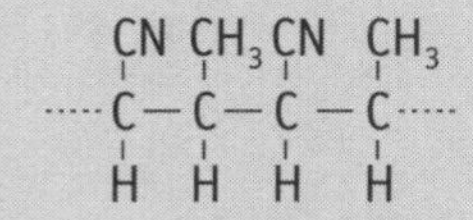

$$\begin{array}{cccc} CN & CH_3 & CN & CH_3 \\ | & | & | & | \\ \cdots C- & C- & C- & C\cdots \\ | & | & | & | \\ H & H & H & H \end{array}$$

   (b) What **pollutants** might be formed by the combustion of this polymer?

The alcohols

In this section the chemistry of this important homologous series of compounds containing the OH group will be considered. You will already have seen how they are named in the 'Naming of organic compounds' section of this Chains and Rings module and you will need to understand about hydrogen bonding, isomerism and fuels.

Background facts

- They are a homologous series containing the **hydroxyl (-OH)** group and their general formula is $C_nH_{2n+1}OH$.
- The most important members of the group are **methanol** (CH_3OH) and **ethanol** (C_2H_5OH) (see later).
- Ethanol is produced on an industrial scale by the reaction of **ethene** with **steam** or by **fermentation** of sugars. ✓ *Quick check 3*
- Alcohols are very soluble in water because they **can form hydrogen bonds** with the water.
- Alcohols burn in air to give CO_2 and H_2O. Because they burn cleanly and quickly they make good fuels. ✓ *Quick check 3*
- The main chemical reactions of alcohols concern the – OH group.

1 Displacement by sodium ✓ *Quick check 1*

Sodium displaces the hydrogen in the –OH group to give hydrogen and sodium alkoxide.

Example: sodium plus ethanol gives hydrogen and sodium ethoxide.

$$2Na(s) + 2C_2H_5OH\ (l) \rightarrow 2C_2H_5O^-Na^+(s) + H_2(g)$$

2 Dehydration (removal of H_2O) using concentrated sulphuric acid (H_2SO_4) or passing over heated pumice (Al_2O_3) gives the alkene. ✓ *Quick check 1*

For example, dehydration of ethanol gives ethene.

$$C_2H_5OH\ (l) \rightarrow H_2O(l) + C_2H_4(g)$$

3 Substitution ✓ *Quick check 1*

The OH group can be **substituted** by bromine using **HBr** (formed from NaBr and conc. H_2SO_4) to give a halogenoalkane.

For example, ethanol and HBr give bromoethane.

$$C_2H_5OH(l) + HBr(g) \rightarrow C_2H_5Br(l) + H_2O(l)$$

4 Esterification

They react with organic acids like ethanoic acid to give **esters**.

For example, ethanol and ethanoic acid react to form the ester, ethyl ethanoate.

$$C_2H_5OH(l) + CH_3COOH(l) \rightarrow CH_3COO\ C_2H_5(l) + H_2O(l)$$

- There are three types of alcohol – primary, secondary and tertiary. These can be distinguished by their reactions with acidified potassium dichromate solution ($H^+/Cr_2O_7^{2-}$).

PRIMARY	SECONDARY	TERTIARY
$CH_3CH_2CH_2CH_2OH$	$CH_3CH(OH)CH_2CH_3$	$CH_3CCH_3(OH)CH_3$
butan-1-ol	butan-2-ol	2-methyl-propan-2-ol

Physical properties of alcohols

✓ Quick check 2

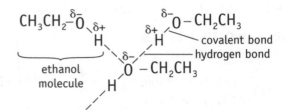

- The alcohol molecule is polar and this leads to hydrogen bonding between the molecules.
- They have higher than expected boiling points because of the hydrogen bonding between the molecules.
- Short chain alcohols are very soluble in water. This is because they can form hydrogen bonds with the water molecules making it energetically favourable to dissolve in the water.
- Alcohols can be recognised from their infrared spectra by the absorption of the O-H bond. This is covered in the section on spectroscopy.

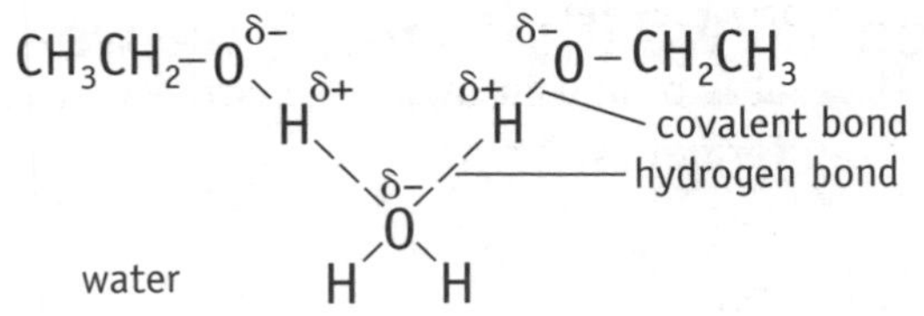

Chemical reactions common to all alcohols

✓ Quick check 1 & 3

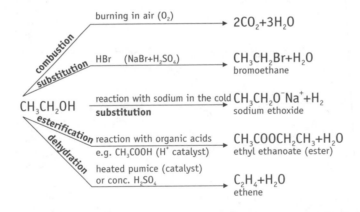

?

Quick check questions

1 Give the **names** and **structural formulae** of the products of the following reactions:

   (a) The dehydration of butan-1-ol.

   (b) The reaction between sodium and butan-1-ol.

   (c) The reaction between 2-methylpropan-2-ol and hydrogen bromide.

2 Draw diagrams to show how methanol hydrogen bonds with:

   (a) itself and (b) water.

3 (a) Write a balanced symbol equation for the combustion of ethanol.

   (b)(i) Explain why ethanol is a good fuel.

   (ii) Why can it be considered to be a renewable fuel?

Types of alcohol

1. **PRIMARY (1°)**. On the carbon with the OH group, there are two hydrogens.

2. **SECONDARY (2°)**. On the carbon with the OH group, there is only one hydrogen.

3. **TERTIARY (3°)**. On the carbon with the OH group there are no hydrogens.

Examples

Primary	Secondary	Tertiary
Butan-1-ol	Butan-2-ol	2-methyl-propan-2-ol
$CH_3CH_2CH_2CH_2OH$	$CH_3CH_2CH(OH)CH_3$	$(CH_3)_3COH$

How do we distinguish between the three types?

- They are distinguished by their ability to be oxidised by an acidified solution of potassium dichromate(VI).

- The **primary** and the **secondary** alcohols each have at least one hydrogen on the carbon with the hydroxyl (OH) group and these can be removed along with the hydrogen on the OH group and these two types are oxidised.

Example 1:

A **primary alcohol**, butan-1-ol. If we heat and distil immediately an **aldehyde** is the organic product.

$$CH_3CH_2CH_2CH_2OH(l) + [O] \rightarrow CH_3CH_2CH_2CHO(l) + H_2O(l)$$

Butanal – an aldehyde

If we use excess potassium dichromate and reflux, then further oxidation can occur to give a **carboxylic acid**, butanoic acid.

$$CH_3CH_2CH_2CH_2OH(l) + 2[O] \rightarrow CH_3CH_2CH_2COOH(l) + H_2O(l)$$

Example 2: A **secondary alcohol**, butan-2-ol. In this reaction a **ketone** is the organic product. Ketones and aldehydes smell very differently and have different chemical reactions.

$$CH_3CH_2CH(OH)CH_3(l) + [O] \rightarrow CH_3CH_2COCH_3(l) + H_2O(l)$$

The secondary alcohol cannot be oxidised further by the acidified dichromate even if the dichromate is in excess.

- The tertiary alcohols have no hydrogens on the carbon with the OH group and therefore cannot be oxidised and there is no reaction. Therefore the colour of the acidified potassium dichromate remains orange.

✓ *Quick check 1*

✓ *Quick check 2*

Note the [O] comes from the oxidising agent, the acidified potassium dichromate.

Ethanol in the form of wine can be oxidised slowly to ethanoic acid in the presence of oxygen.

✓ *Quick check 4*

The industrial production of ethanol

Because ethanol has several important uses it is necessary to prepare it on a large scale. There are two main methods and they are very different.

Method 1 – The fermentation of sugars using yeast

- Juices of fruits, which contain sugars are fermented in the presence of yeast.
- The yeast contains enzymes which catalyse the reactions.
- The overall equation is:

$$C_6H_{12}O_6(aq) \rightarrow 2C_2H_5OH(aq) + 2CO_2$$

- When the ethanol concentration reaches 12% then the yeast are killed and the reaction stops.
- The ethanol produced in this way has a taste dependant on the source of the fruit from which it was made.

Method 2 – The hydration of ethene

- In this method, ethene and steam are passed at 300°C and a high pressure over a catalyst of phosphoric acid on silica pellets.

$$C_2H_4(g) + H_2O(g) \rightarrow C_2H_5OH(g)$$

- This alcohol is industrial alcohol and is not used as a beverage.

The main uses of alcohols

Alcohol	Uses
Methanol	In the manufacture of thermosetting plastics such as phenol-formaldehyde resins (Bakelite). In the manufacture of Perspex.
Ethanol	As a fuel for cars in countries with no oil reserves. As a solvent for several organic chemicals. As a beverage (in wines, beers and sprits). In the preparation of esters.

✓ **Quick check 3**

Countries such as Brazil with few oil reserves obtain alcohol from sugar. It is made more concentrated and then used as a fuel for cars. Because the source of the ethanol is sugar and this can be grown again and again. Therefore ethanol is a renewable fuel.

? Quick check questions

1 Classify the following alcohols as primary, secondary or tertiary.

(a) $CH_3CH_2CH_2OH$

(b) $CH_3CH_2CH(OH)CH_2CH_3$

(c) $C_6H_5CH_2OH$

(d) $(CH_3)_3COH$

(e) $C_6H_5CH(OH)CH_3$

2 Explain how you can distinguish between 2-methylpropan-2-ol and butan-2-ol.

3 Explain why production of alcohol by fermentation can never produce 100% alcohol straight away.

4 Explain why when wine is left opened for any length of time it starts to taste of vinegar (vinegar is ethanoic acid – a carboxylic acid).

Infrared spectroscopy

Infrared spectroscopy is a very useful technique in the determination of a compound's structure. It relies on the fact that different bonds in compounds absorb at different frequencies in the infrared region of the spectrum. In infrared analysis the **wavenumber** (the reciprocal of wavelength) are used instead of frequencies or wavelengths and are measured in cm^{-1}. The functional groups and bonds you will be asked about are given in the table below along with their **wavenumbers**. Note that you do not have to remember these in the exam, as they will be given in your databook.

Functional group/bond	Wavenumber (cm^{-1})
Hydroxyl (-OH) in carboxylic acids	Very broad 2500 to 3300
O-H 'hydrogen bonded' in alcohols and phenols	Less broad between 3230 to 3550
Carbonyl group(C=O) in aldehydes, ketones and carboxylic acids	1680–1750

Some typical spectra for compounds exhibiting these infrared absorptions are shown below/ on the opposite page.

Note the following:

I   The alcohol shows a strong broad absorption at about 3400 cm^{-1} due to the O–H hydrogen bonding and no absorption at 1740 cm^{-1} because there is no carbonyl group (C=O).

II The carboxylic acid shows a very broad absorption at around 2500–3000 cm^{-1} because of the of the O-H and at approximately 1700 cm^{-1} due to the carbonyl group.

III The aldehyde and the ketone show an absorption at around 1700 cm^{-1} but no strong broad absorption at ~3000 cm^{-1}.

IV For the alcohols, esters and carboxylic acids there is a C–O absorption between 1000–1300 cm^{-1} but this is harder to see.

? Quick check questions

1   A compound, X, gives a strong, broad absorption at around 3300–3400 cm^{-1} but none at 1700 cm^{-1}.

    (a) Which functional group is present in X ?

    (b) To which homologous series does it belong?

2   Describe what changes you would see in the infrared spectra when ethanol (CH_3CH_2OH) oxidised to ethanal (CH_3CHO) and then to ethanoic acid (CH_3COOH).

3   Ethanol is isomeric with a substance called ethoxyethane (CH_3OCH_3). Explain how the two compounds can be distinguished using infrared spectroscopy.

4   Explain how you can distinguish between propan-2-ol ($CH_3CH(OH)CH_3$) and propanoic acid (CH_3CH_2COOH).

The areas of the spectra that are of main interest have been enclosed by rectangles.

I   Alcohol

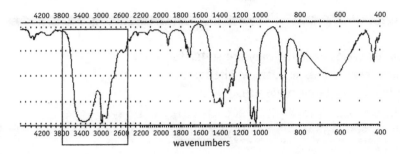

wavenumbers

II   Carboxylic acid

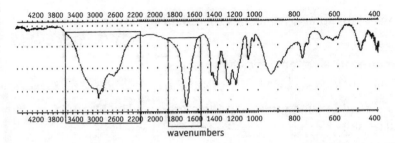

wavenumbers

III   Carbonyl-aldehyde

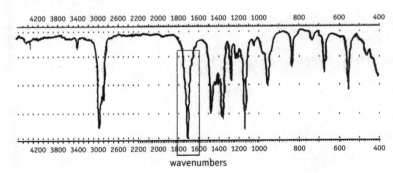

wavenumbers

IV   Carbonyl-ketone

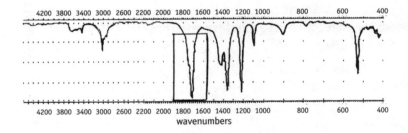

wavenumbers

The halogenoalkanes

The halogenoalkanes are an important homologous series because they are so useful as intermediates in the synthesis of other organic compounds especially when a longer carbon chain is needed. They also have important large-scale uses.

Background facts

- They all contain at least one halogen (abbreviated to hal) attached to a carbon atom (C–hal).

- If they contain just one halogen atom and there are no multiple bonds then their general formula is C_nH_{2n+1}hal.

✓ *Quick check 1*

- The C–hal bond is **polarised** leaving the carbon atom electron deficient and open to attack by **nucleophiles** [e.g. OH^- ions from (KOH(aq) and ammonia (NH_3)].

✓ *Quick check 3*

> **Remember** A nucleophile is a molecule or ion, which can donate a pair of electrons. A nucleophile is quite often represented as $Nu:^-$

- Because the halide ion (hal^-) is stable, it is **a good leaving group** and therefore it is substituted reasonably easily by the nucleophile.

- The reaction mechanism undergone by the halogenoalkanes when they are substituted is **nucleophilic substitution.** This literally means **substitution** by a **nucleophile**.

- The best test for halogenoalkanes is to warm with water and then test for the halide ion using **aqueous silver nitrate solution** (see page 62).

- There are three types of halogenoalkanes – **primary**, **secondary** and **tertiary.**

> The strength of the C–hal bond is in the following order (strongest first):
>
> ### C-F > C-Cl > C-Br > C–I
>
> The reactivity of the halogenoalkanes is in the reverse order:
>
> MOST REACTIVE $\rightarrow$ LEAST REACTIVE
>
> C-I >   C-Br >   C-Cl >   C-F

✓ *Quick check 4*

The reactivity of the halogenoalkanes depends on the breaking of the C–hal bond. Since the C–F bond is the strongest it is therefore the hardest to break and therefore fluoroalkanes are the least reactive of the halogenoalkanes.

- In the presence of concentrated ethanolic alkali (a solution of potassium hydroxide in ethanol), they lose a hydrogen halide molecule (H–hal) – they undergo elimination to form an alkene.

- Halogenoalkanes, especially those containing chlorine and fluorine, have several uses related to their properties.

- **Chlorofluorocarbons** (CFCs) which were once used as refrigerants and in fire extinguishers and are very stable and persist in the environment. They catalyse the decomposition of ozone, leading to holes in the ozone layer.

- Unsaturated halogenoalkanes such as chloroethene ($CHCl=CH_2$) and tetrafluoroethene ($CF_2=CF_2$) act just like normal alkenes and can be used to make polymers.

Types of halogenoalkane

✓ *Quick check 2*

1. **PRIMARY (1°).** On the carbon with the halogen, there are two hydrogens.

2. **SECONDARY (2°).** On the carbon with the halogen, there is only one hydrogen.

3. **TERTIARY (3°).** On the carbon with the halogen there are no hydrogens.

Examples

Primary	Secondary	Tertiary
1-iodobutane	2-chlorobutane	2-bromo-2-methylpropane
$CH_3CH_2CH_2CH_2I$	$CH_3CH_2CHClCH_3$	$(CH_3)_3CBr$

Reactions of halogenoalkanes

- The electron-deficient carbon in the carbon–halogen bond is liable to attack from nucleophiles with their lone pair electrons.

- The carbon cannot form more than four bonds so the nucleophile (Nu:⁻) forms a covalent bond with the carbon and the halogen leaves as a hal⁻ (e.g. Br⁻) ion.

The general mechanism for bromoethane and a nucleophile (Nu:⁻) is shown below:

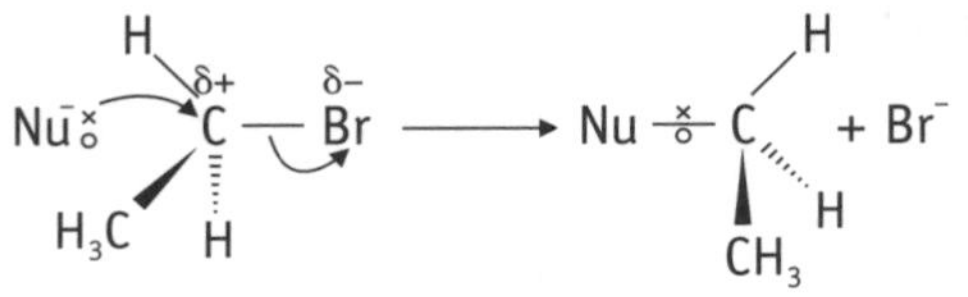

The nucleophile approaches the electron-deficient carbon and using its lone-pair electrons forms a dative covalent bond with the carbon.

The Br⁻ is stable and is able to leave the bromoethane. The nucleophile has replaced (substituted) it.

Quick check questions

1 Analysis of a halogenoalkane, A showed that it contained 29.3% carbon, 5.7% hydrogen and 65% bromine. (relative atomic masses Br = 80; C = 12; H = 1)

   (a) Calculate the empirical formula of A.

   (b) Its relative molecular mass is 123 g.mol⁻¹. Calculate its molecular formula.

   (c) Draw the displayed formulae for the isomers of A.

2 Give the isomers of bromobutane (C_4H_9Br) and classify them as 1°, 2° or 3°.

3 Explain how the following substances can act as nucleophiles.

   (a) NH_3 (b) KCN   (c) KOH

4 (a) When halogenoalkanes are used for synthesis of carbon compounds, fluoroalkanes are never used. Explain why.

   (b) Suggest a reason why bromoalkanes are used in preference to iodoalkanes.

Reactions of the halogenoalkanes

Halogenoalkanes, especially those with C–I and C–Br bonds are useful intermediates in synthetic pathways as other groups can be introduced into the molecule easily. Bromoalkanes are probably the most useful.

Example - **Bromoethane**

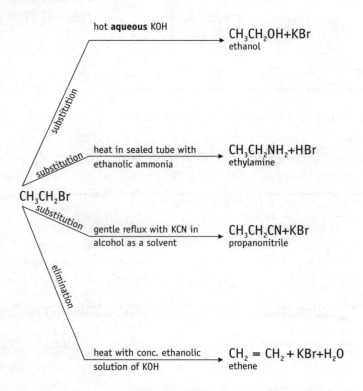

▶ The nucleophile here is the OH⁻ (aq) ion.

▶ The nucleophile here is the NH_3 molecule.

▶ The nucleophile here is the CN⁻ ion. A good way to lengthen the carbon chain. Very important in synthesis of carbon compounds.

▶ This is an elimination reaction. The molecule of bromoethane loses a molecule of HBr.

Comparing the reactivities of the halogenoalkanes

An experiment to compare their reactivities:

1 In three separate test tubes, mix equal volumes of $AgNO_3$ solution in alcohol and warm to 40°C.

2 Add a few drops of 3 comparable halogenoalkanes (e.g. 1-chlorobutane, 1-bromobutane and 1-iodobutane) to each tube. The first reaction is:

R–hal(l) + H_2O(l) → ROH(l) + H⁺(aq) + hal⁻ (aq)

Silver ions (from $AgNO_3$) then react with the halide (hal⁻) ions giving a precipitate of silver halide:

Ag⁺(aq) + hal⁻ (aq) → Aghal(s)

This precipitate will make the mixture go cloudy.

3. The speed with which the mixture goes cloudy is an indication of the speed of the reaction and hence the reactivity of the C–halogen bond.

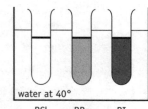

water at 40°
RCl   RBr   RI
with AgNO₃ (aq) in alcohol as a solvent
(R is an alkyl group)

Uses of halogenoalkanes

The C–F and C–Cl bonds are strong and therefore compounds containing these are relatively unreactive and stable under normal atmospheric conditions. Of course in the presence of a nucleophile they are reactive (see reactions of halogenoalkanes). If a compound contains both chlorine and fluorine they are called: **CHLOROFLUOROCARBONS (CFCs)** and **until recently*** they had several uses. Since it was discovered that CFCs damage the ozone layer their use has been discontinued and chemists have had to find alternatives.

Use	Properties and reason(s) for use
Refrigerants	Unreactive - therefore will not corrode machinery parts. Volatile - so easily vaporised.
Dry cleaning	Good solvents – will dissolve grease.
Aerosols	Volatile - so easily vaporized along with the other contents of the aerosol can.
Fire-extinguishers	Unreactive and do not burn easily.

✓ *Quick check 4*

*The use of CFCs has been discontinued because of their effect on the ozone

? Quick check questions

1 What are the **three** isomeric products of the reaction of 2-bromobutane with ethanolic KOH?

2 Give a full balanced equation for the reactions between 1,2-dibromoethane and the following:

(a) KOH (aq)

(b) KCN

(c) NH_3

3 Explain how you could get from bromoethane to polyethene in 2 steps.

4 Explain why chlorofluoroalkanes had the following uses:

(a) Refrigerants

(b) Fire extinguishers,

(c) Dry-cleaning

5 Explain how you could show that the C—Br bond is more easily broken than the C–Cl bond.

Yields in organic reactions

All organic chemists need to know the efficiency of their preparations. The most frequent question asked in this area of the syllabus is, 'what is the percentage yield?'

Put simply the percentage yield may be expressed as:

$$\text{PERCENTAGE YIELD} = \frac{\text{'What you get'}}{\text{'What you should get'}} \times 100\%$$

Learn the formula for working out the percentage yield

Worked example

When 9.2 g of ethanol (M_r=46) is reacted with excess hydrogen bromide (from NaBr and concentrated H_2SO_4), 13.08 g of bromoethane (M_r=109) are formed. What is the percentage yield?

Step 1: We have to find out **what we should get** and first we have to construct the equation for the reaction. $HBr + C_2H_5OH \rightarrow C_2H_5Br + H_2O$

Step 2: Calculate the number of moles of ethanol in the reaction.

$$\text{No. of moles} = \frac{\text{mass}}{M_R} = 9.2/46 = 0.2 \text{ mol}$$

Step 3: From the equation see that in this reaction 1 ethanol gives 1 bromoethane and therefore 0.2 mol. **should** give 0.2 mol. of bromoethane.

Step 4: Calculate the mass of bromoethane **we should get.**

Mass of bromoethane = no. of mol x M_R = 0.2 x 109g = 21.8 g

Step 5: Calculate the percentage yield.

$$\frac{\text{'What you get'}}{\text{'What you should get'}} \times 100\% = \frac{13.08}{21.8} \times 100\% = \underline{60\%}$$

Another question you might be asked about yields is **'Suggest why the percentage yield is substantially below 100%'.**

The **wrong** answer to this question is anything suggesting human error.

Examples:
1 The student weighed it out wrongly.
2 The student dropped some of it.

Your answers to this type of question should describe something about the reaction.

Examples:
1 When distilling not all the distillate was evaporated from the mixture.
2 When recrystallising not all of the crystals crystallised from solution.
3 The reaction did not go to completion.
4 The receiver was not cooled therefore not all the distillate condensed.

Quick check questions

1 When 18.0 g of propan-1-ol (C_3H_8O) was reacted with excess acidified sodium dichromate, 7.20 g of propanal (C_3H_6O) were obtained. Calculate the percentage yield given that the equation for the reaction is:
$3C_3H_8O + Na_2Cr_2O_7 + 4H_2SO_4 \rightarrow 3C_3H_6O + Na_2SO_4 + Cr_2(SO_4)_3 + 7H_2O$

2 In the preparation of bromoethane, ethanol (CH_3CH_2OH) is reacted with HBr according to the following equation:
$CH_3CH_2OH + HBr \rightarrow CH_3CH_2Br + H_2O$
In a preparation, 9.2 g of ethanol gave 5.36 g of bromoethane. What is the percentage yield?

Module B: End-of module questions

1 The diagrams below show the structures of four isomers of molecular formula C_4H_8.

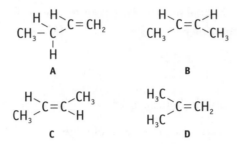

A   B

C   D

(a)(i)   To which class of compounds do the four isomers belong?

(ii)   Which two diagrams show compounds which are *cis-trans* isomers?   [2]

(b)(i)   Which of these compounds could be formed from 2-methylpropan-2-ol by the elimination of water?

(ii)   State the reagents and conditions by which this reaction could be carried out in the laboratory   [3]

2 Samples of butan-1-ol, butan-2-ol and 2-methylpropan-2-ol were reacted with aqueous acidified potassium dichromate (VI).

Complete the table below for each of the three alcohols. State what you would observe and identify the organic product of each reaction.   [5]

alcohol	observation	organic product (if any)
butan-1-ol		
butan-2-al		
2-methylpropan-2-ol		

3 An important use of chlorine is in the production of plastics such as polychloroethene, i.e. PVC.

The monomer is shown below.

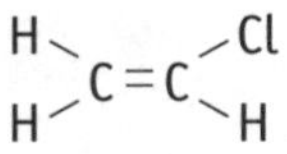

$$H_2C=CHCl$$

monomer

(a) Draw the displayed formula for the repeating unit of PVC.   [1]

(b) State TWO problems that arise from the disposal of PVC.   [2]

4 Explain the following reaction terms as used in organic chemistry. In each case give **one** example reaction. For each example name the reactants, state the conditions, name the organic product(s) and give an equation.

(a) Dehydration (b) Elimination (c) Nucleophilic substitution.   [12]

5 Cracking of the unbranched compound **E**, C_6H_{14}, produced the saturated compound **F** and an unsaturated hydrocarbon **G** (Mr 42). Compound **E** reacted with bromine in UV light to form a monobrominated compound **H** and an acidic gas **I**. Compound **G** reacted with hydrogen bromide to form a mixture of two compounds **J** and **K**.

(a) Use this evidence to suggest the identity of each of compounds **E** to **K**. Include equations for the reactions in your answer. [10]

(b) Oil companies often 'reform' compounds such as **E**. Explain why this is done and **suggest two** organic products of the reforming of **E**. [3]

(c) Predict the structure of the polymer that could be formed from compound **G**. [1]

6. Petrol and camping gas are examples of fuels that contain hydrocarbons.

(a) Petrol is a mixture of alkanes containing 6 to 10 carbon atoms per molecule. Some of these alkanes are isomers of one another.

   (i) Explain the term *isomers*.

   (ii) State the molecular formula of an alkane present in petrol. [2]

(b) The major hydrocarbon in camping gas is butane. Some camping gas was reacted with chlorine to form a mixture of isomers.

   (i) What conditions are required for this reaction to take place?

   (ii) Two isomers, **A** and **B**, were separated from this mixture. These isomers had a molar mass of 92.5 g mol^{-1}.

   Deduce the molecular formula of these two isomers

   (iii) Draw the displayed formulae of **A** and **B** and name each compound. [6]

Module C: How Far, How Fast?

How Far, How Fast? is a half module unit, so contains half as much material as the Foundation and Chains and Rings modules. It deals with physical chemistry – explaining chemical behaviour by describing the rules that govern it. The content is split into 3 sections, Enthalpy Changes, Reaction Rates and Chemical Equilibria.

Topic	Reference	Previous knowledge required
Enthalpy Changes Exothermic and endothermic changes Enthalpy changes, ΔH Standard conditions Enthalpy profile diagrams Enthalpy change of reaction, formation and combustion Bond enthalpy Hess's Law Fuel combustion and photosysthesis	5.3.1	GCSE material on exothermic and endothermic changes. Heat loss or gain in a reaction. Using simple mathermatical equations. You may have met these in GCSE work. Combustion reactions – Foundation module and GCSE work Writing balanced chemical equations – Foundation module. Maths skills are used here, especially rearranging equations. Useful products from oil – petrol and other fuels.
Reaction Rates Collision theory Effect of concentration and temperature on reaction rate Activation energy Catalysts – homogeneous and heterogeneous Carbon monoxide, nitrogen oxides and hydrocarbons as pollutants Catalytic converters	5.3.2	Particulate nature of matter, and how the particles behave in solids, liquids and gases. GCSE work on rates of reactions, and how temperature affects rate. GCSE work on catalysts. Pollution from car engines. GCSE work on carbon chemistry. Combustion of alkanes from Chains and Rings.
Chemical Equilibria Reversible reactions Dynamic equilibrium Le Chaterlier's principle The Haber process Acids and bases Conjugate acids and bases Strong and weak acids Commercial importance of ammonia-derived compounds	5.3.3	GCSE work on $NH_4Cl \rightleftharpoons NH_3 + HCl$. GCSE work on Haber process. GCSE work on acids, alkalis and bases. GCSE work on fertilisers and ammonia.

Enthalpy changes

Chemical reactions can be divided into two types – those which cause a rise in temperature and those which cause a fall in temperature. We have special names for these types of reaction:

exothermic reactions	*heat is given out*, the temperature *increases*
endotahermic reactions	*heat is taken in*, the temperature *decreases*

Most chemical changes are accompanied by a change in temperature. There is a particular name for a heat change that takes place at *constant pressure*, such as normal laboratory conditions – it is called an **enthalpy change**.

Enthalpy is given the symbol H, and an **enthalpy change** is given the symbol ΔH.

The unit for H and ΔH is the joule, **J**.

Often ΔH is given in **kilojoules, kJ**, which is $J \times 10^{-3}$.

$$\Delta H = H_{products} - H_{reactants}$$

> ◗ Δ is the symbol *delta*. It is the symbol for the difference between two values

> ◗ 1 kJ = 1000 J
> 1 J = 1 kJ ÷ 1000

Enthalpy profile diagrams

A chemical reaction can be regarded as *breaking bonds* followed by *making bonds*. Breaking bonds needs energy, so energy must be put in to the reaction. This energy is called the **activation energy**. After that, bonds are made and this releases energy.

An **exothermic reaction** gives out more energy making bonds than it takes in breaking bonds so we see a temperature rise.

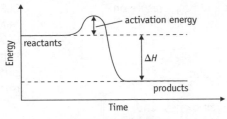

EXOTHERMIC REACTION
reaction has lost energy so ΔH is negative

An **endothermic reaction** needs more energy than it gives out so we see a temperature fall.

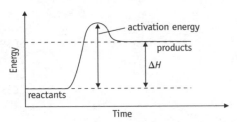

ENDOTHERMIC REACTION
reaction has gained energy so ΔH is positive

H	the symbol for **enthalpy** measured in **J** or **kJ**
ΔH	the symbol for the **enthalpy change in a reaction** measured in **J** or **kJ**
activation energy	the minimum energy reactants need before they can react
exothermic reaction	ΔH **is negative** temperature rises
endothermic reaction	ΔH **is positive** temperature falls

How to calculate enthalpy changes

For reactions involving solutions, use this equation to calculate the enthalpy change of a reaction:

$$\Delta H = m \times c \times \Delta T$$

What do these symbols mean?

- ΔH is the enthalpy change you are calculating. It must always have a sign, + or -, depending on whether it is an exothermic reaction (-) or an endothermic reaction (+). The units are usually **kJ** or **kJ mol^{-1}**.
- m is the mass, but in solution we can assume that the mass is the same as the volume, as 1 g of water has a volume of 1 cm^3. This means that m **is the final volume in cm^3.**
- c is a **constant** which is the *heat capacity of the **calorimeter***. A calorimeter is a container an enthalpy change is measured in – it is insulated so no heat escapes. c is **4.2 Jg^{-1}K^{-1}**
- ΔT is the difference in temperature from the start of the reaction to the highest or lowest temperature reached.

> Don't put a sign, + or −, in front of ΔT in this equation!

Worked example

50 cm^3 of concentrated sulphuric acid, 20 mol dm^{-3}, was added to 950 cm^3 of water at a temperature of 19°C. The solution was stirred gently until the maximum temperature was reached at 28°C. What is the enthalphy change for this reaction?

Step 1: Write down the equation you will use: $\Delta H = mc\Delta T$

Step 2: Write down the values for the symbols m, c and ΔT:

$m = 950 + 50 = 1000$ cm^3; $c = 4.2$; $\Delta T = 28 - 19 = 9$°C.

Step 3: Do the calculation: $\Delta H = 1000 \times 4.2 \times 9 = 37800$ J

Step 4: To get the answer in kJ, divide by 10^3:

$37800 \div 10^3 = 38$ kJ (you can round this to two sig. fig.)

Step 5: Now put the *sign* in. Look at the temperature change – it goes *up*. This is an *exothermic* reaction so it has a *negative* sign: **−38 kJ**

> c is always 4.2 if the volume is in cm^3 and the temperature is in °C

How to work out the enthalpy change per mole

Enthalpy changes are often given in kJ mol^{-1}. This means the energy in kJ given out or taken in for every mole. To get this, *divide the enthalpy change you have calculated by the number of moles*.

Worked example

What is the enthalpy change per mole of sulphuric acid in the example above?

Step 1: Calculate the enthalpy change for the reaction : **−38 kJ**

Step 2: Calculate the number of moles of sulphuric acid:

$$\text{no. moles} = \frac{\text{vol. in cm}^3}{1000} \times \text{concn} = 50/1000 \times 20 = 1.0 \text{ mol}$$

Step 3: Divide the ΔH value by the number of moles: $38 \div 1.0 = -38$ kJmol^{-1}

? Quick check question

1 A solution of ammonium nitrate was made by adding 0.5 mol NH_4NO_3 to water at a temperature of 19°C to make a final volume of 500 cm^3. The temperature fell to 12.8°C. Calculate the enthalpy change per mole of ammonium nitrate dissolving.

Enthalpy change definitions

Now you have to learn five enthalpy change definitions:

(1) The Enthalpy Change of Reaction ΔH is the heat exchange with the surroundings at constant pressure.

This means just the enthalpy change for the reaction as it stands, no matter how many moles are involved, calculated using $mc\Delta T$.

(2) The Standard Enthalpy Change of Reaction $\Delta H^{\ominus}$ is the enthalpy change for the given reaction *under standard conditions*.

(3) The Standard Enthalpy Change of Formation $\Delta H_f^{\ominus}$ is the enthalpy change when 1 mole of the compound is formed from its elements under standard conditions.

> Standard conditions are:
> pressure 101 kPa; temperature 298K; solutions 1 mol dm^{-3}

> A standard enthalpy change has the symbol $\Delta H^{\ominus}$; the $\ominus$ means standard conditions.

> **ΔH_f for an element is zero**

Example 1: $\Delta H_f^{\ominus}$ (CH_4) is the enthalpy change for the formation of 1 mole of methane. The equation for this is $C(s) + 2H_2(g) \rightarrow CH_4(g)$.

State symbols are important here. Methane is made up of carbon and hydrogen, which as natural elements are solid carbon and gaseous hydrogen. Methane is a gas.

Example 2: $\Delta H_f^{\ominus}$ (MgO) – the equation is $Mg(s) + \frac{1}{2}O_2(g) \rightarrow MgO$. The equation must be balanced in this way because the enthalpy change of formation is for making *1 mole* of *compound*.

(4) The Standard Enthalpy Change of Combustion $\Delta H_c^{\ominus}$ is the enthalpy change when 1 mole of the substance is burned completely under standard conditions.

Example 1: $\Delta H_c^{\ominus}$ (C(s)) is the enthalpy change when 1 mole of carbon is completely burnt. The equation is $C(s) + O_2(g) \rightarrow CO_2(g)$.

Example 2: $\Delta H_c^{\ominus}$ ($CH_4(g)$): The equation is $CH_4(g) + 2O_2(g) \rightarrow CO_2(g) + 2H_2O(l)$. Each type of atom in the substance being burnt forms a compound with oxygen. The equation must be balanced to give 1 mole of the substance because *by definition* the enthalpy change of combustion is for *burning 1 mole*.

(5) The Standard Enthalpy Change of Bond Dissociation $\Delta H_d^{\ominus}$ **(often called the Bond Enthalpy)** is the enthalpy change when 1 mole of bonds of a particular type are broken under standard conditions. IT IS ALWAYS POSITIVE – breaking bonds needs energy, making bonds releases energy.

Example: $\Delta H_d^{\ominus}$ (HCl) is the enthalpy change when 1 mole of H–Cl bonds are broken: $HCl(g) \rightarrow H(g) + Cl(g)$.

A bond enthalpy for a particular bond is actually an approximate value, because the strength of the bond depends on the rest of the molecule; the C–H bond in CH_4 is not quite the same as the C–H bond in CH_3OH.

Hess's Law

> Hess's Law states that the enthalpy change for any reaction is always the same, no matter what route is used.

It is a law of *energy conservation*. In this reaction

A+B can make C+D in 1 Step (ΔH_1) ... or 2 Steps $(\Delta H_2 + \Delta H_3)$.

$\Delta H_1 = \Delta H_2 + \Delta H_3$

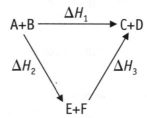

To work out this equation, you need to find the *start* and the *finish* of the Hess cycle.

start – both arrows diverge

finish – both arrows converge

We use Hess's law to find an enthalpy change which is difficult to find experimentally:

Enthalpy change of formation from enthalpy changes of combustion

ΔH_f of a compound = ΔH_c of the elements $-$ ΔH_c of the compound

$$\text{elements} \xrightarrow{\Delta H_f^{\ominus}} \text{compound}$$

combustion products

Example: $C(s) + 2H_2(g) \rightarrow CH_4(g)$

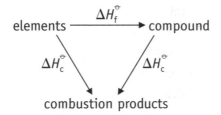

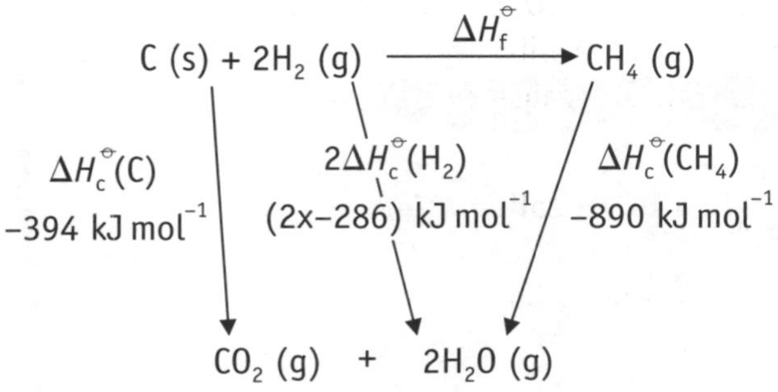

$$\Delta H_c^{\ominus}(C) \quad -394\ \text{kJ mol}^{-1}$$

$$2\Delta H_c^{\ominus}(H_2) \quad (2\times-286)\ \text{kJ mol}^{-1}$$

$$\Delta H_c^{\ominus}(CH_4) \quad -890\ \text{kJ mol}^{-1}$$

$$CO_2(g) + 2H_2O(g)$$

$$\Delta H_f^{\ominus}(CH_4) = \Delta H_c^{\ominus}(C) + 2\Delta H_c^{\ominus}(H_2) - \Delta H_c^{\ominus}(CH_4)$$

$$= (-394) + (2\times-286) - (-890)$$

$$= -394 - 572 + 890$$

$$\Delta H_f^{\ominus}(CH_4) = -76\ \text{kJ mol}^{-1}$$

Note that ΔH_c values are for burning 1 mole, so you must remember to multiply them appropriately if more than 1 mole is burnt.

Enthalpy change of reaction from enthalpy changes of formation

$\Delta H_{reaction}$ = ΔH_f of the products − ΔH_f of the reactants

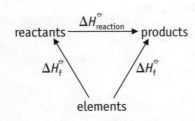

Example:

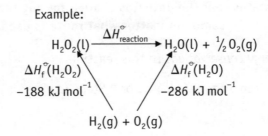

$$\Delta H^{\ominus}_{reaction} = \Delta H^{\ominus}_f (H_2O) - \Delta H^{\ominus}_f (H_2O_2)$$

$$= (-286) - (-188)$$

$$\Delta H^{\ominus}_{reaction} = -98 \text{ kJ mol}^{-1}$$

> Remember, ΔH_f for an element is zero!

Enthalpy change of a reaction from bond enthalpies

$\Delta H_{reaction}$ = ΔH_d of the products   − ΔH_d of the reactants

$\Delta H_{reaction}$ = BE of all the bonds in the reactants − BE of all the bonds in the products

CH_4 (g)	+	Br_2 (g)	→	CH_3Br (g)	+	HBr (g)
H H−C−H H		Br−Br		H H−C−Br H		H−Br
4(C–H) bonds		1(Br–Br) bond		3(C–H) bonds 1(C–Br) bond		1(H–Br) bond
4x(413)		1x(193)		3x(413) + 1x(290)		1x(366)
1652		193		1529		366

$$\Delta H_{reaction} = (1529 + 366) - (1652 + 193)$$

$$= 1895 - 1845$$

$$\Delta H_{reaction} = +50 \text{ kJ mol}^{-1}$$

? Quick check questions

1 Calculate the enthalpy of formation of methane given:

$CH_4(g) + 2O_2(g) \rightarrow CO_2(g) + 2H_2O(l)$ $\Delta H = -890.4$ kJ mol^{-1}

$\Delta H^{\ominus}_f (CO_2)(g)) = -394$ kJ mol^{-1}; $\Delta H^{\ominus}_f (H_2O)(l)) = -286$ kJ mol^{-1}

2 Calculate the standard enthalpy of formation of ammonia given that:

$\Delta H^{\ominus}_c (NH_3)(g)) = -286.5$ kJ mol^{-1}; $\Delta H^{\ominus}_f (NO_2)(g)) = +33.4$ kJ mol^{-1};

$\Delta H^{\ominus}_f (H_2O)(l)) = -242$ kJ mol^{-1}

Reaction rates

Some reactions are fast, some are slow – you will study how reaction rates are influenced by temperature and catalysts.

The collision theory

This theory states that:

- particles must *collide* in order to react
- for a successful reaction, the collision must be *energetic enough* to overcome the **activation energy** (the minimum energy needed)

✓ *Quick check 1*

- for a successful reaction, the particles must be in *correct orientation* to each other.

We use the collision theory to explain

1 the effect of concentration changes on reaction rate

Fact: an increase in the concentration of a solution or the pressure of a gas increases the reaction rate.

Explanation: if the concentration of a solution or the pressure of a gas is increased, then there are more particles present. This means there will be more collisions so a greater chance of a collision overcoming the activation energy and giving a reaction.

2 the effect of temperature on reaction rate

Fact: a increase in temperature increases the reaction rate.

Explanation: if the temperature is increased, the particles move faster. This means there will be more collisions and so a greater chance of a collision overcoming the activation energy and giving a reaction. This is also shown by:

The Boltzmann distribution

✓Quick check 3

In a sample of gas, the particles have different speeds. This means they also have different energies. The Boltzmann distribution shows us the *distribution of the molecular energies in a sample of gas.* (A)

If the temperature is increased, the distribution alters – the maximimum of the curve is lower and moves to the right. This increases the number of molecules which have the activation energy, so the reaction rate increases. (B)

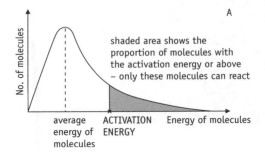

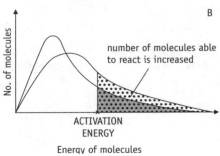

Catalysts

A catalyst is a substance that enters into the reaction and increases the reaction rate without being consumed. It provides an alternative reaction pathway of *lower activation energy*.

Catalysts have great economic importance. They are used in three major areas: fertiliser production, margarine production and petroleum processing.

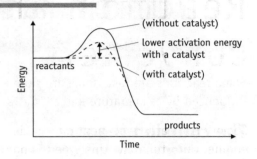

Homogeneous catalysts

These are catalysts which are in the *same phase* as the reaction mixture.

Example 1: H^+ as a catalyst in the reaction *carboxylic acid + alcohol → ester + water*

Here the catalyst is an acid such as sulphuric acid which is added to the reaction mixture. All the reactants and products are in solution, just like the acid, so this is a *homogeneous catalyst*.

Example 2: chlorine free radicals in the breakdown of ozone to oxygen, where all the species are gases:

$$Cl\bullet + O_3 \rightarrow ClO\bullet + O_2$$

$$ClO\bullet + O \rightarrow Cl\bullet + O_2$$

This reaction shows how the hole in the ozone layer is formed. It is a chain reaction because the chlorine free radical is regenerated, so can react with more ozone. The O in the equation means oxygen atoms which are formed in the stratosphere.

Heterogeneous catalysts

These are catalysts which are in a *different phase* as the reaction mixture.

Example 1: iron as a catalyst in the Haber process (see page 78).

The Haber process is the reaction between the gases nitrogen and hydrogen to give ammonia

$$N_2 \text{ (g)} + 3H_2 \text{ (g)} \rightarrow 2NH_3 \text{ (g)}$$

Iron is a solid and everthing else is gaseous, so this is a heterogeneous catalyst.

Example 2: a mixture of platinum, palladium and rhodium in catalytic converters (see page 74).

✓ *Quick check 2*

? *Quick check questions*

1 Sketch an enthalpy change diagram for the combustion of methane, labelling the reactants, products and enthalpy change.

2 Aqueous potassium dichromate is frequently used to catalyse the oxidation of aldehydes. Is this a homogeneous or heterogeneous catalyst?

3 Draw a Boltzmann distribution of gas molecules and use it to show that raising the temperature increases the rate of reaction.

Pollution and catalytic converters

Society is greatly dependent on vehicles powered by the internal combustion engine. Unfortunately, this type of engine is also a major pollutor. **Catalytic converters** help to remove pollutants from exhaust gases using a solid catalyst. They consist of a cylindrical structure containing a *ceramic honeycomb* coated with the catalyst – a mixture of *platinum, palladium and rhodium* (see page 44). The honeycomb structure *increases the surface area* of the catalyst so it is more effective. Hot exhaust gases are mixed with air and pass through the catalytic converter, which is fixed to the exhaust.

It is worth noting that the catalytic converter converts hydrocarbons and carbon monoxide (CO) to carbon dioxide (CO_2). This is not a complete answer to the pollution problem, as CO_2 is a *greenhouse gas* which accumulates in the upper atmosphere and prevents heat loss from the earth. This causes *global warming*.

Pollutant	Hydrocarbons	Carbon monoxide CO	Oxides of nitrogen NO, NO_2, N_2O etc
Where does it come from?	Not all fuel (petrol or diesel) is burnt in the engine. Some escapes unchanged.	CO is formed when hydrocarbons burn inefficiently, with not enough oxygen. This always happens to some extent.	Nitrogen and oxygen in the air combine in the heat of the engine to form nitrogen oxides.
What effect does the pollutant have?	Hydrocarbons are toxic and carcinogenic.	Carbon monoxide is highly toxic. It combines with haemoglobin in the blood more efficiently than oxygen, and so prevents oxygen reaching the tissues.	Nitrogen dioxides have 3 main effects: (1) They cause *smog*, a yellowish haze which is corrosive and irritating to the respiratory tract. It has been blamed for an increase in the incidence of asthma. (2) They cause low-level *ozone*, also found in smog. (3) They contribute to *acid rain*.
What does the catalytic converter do to solve the problem?	It oxidises hydrocarbons to carbon dioxide and water.	It oxidises carbon monoxide to carbon dioxide and water **step 1:** adsorption of CO onto active sites on the catalyst surface **step 2:** chemical reaction **step 3:** desorption of CO_2 from the catalyst surface	It reduces nitrogen oxides to nitrogen, N_2. **step 1:** adsorption of NO onto active sites on the catalyst surface **step 2:** chemical reaction **step 3:** desorption of N_2 from the catalyst surface

? Quick check question

1 Vehicle exhausts emit the pollutant nitrogen dioxide. Describe *one* effect of this pollutant, and explain how its emission can be controlled.

Chemical equilibrium

An **equilibrium** is two opposing processes taking place at *equal rates*.

We use the symbol $\rightleftharpoons$ to show this: $2NO_2$ (g) $\rightleftharpoons$ N_2O_4 (g) means that when a **steady state** has been reached, nitrogen dioxide NO_2 is constantly changing to dinitrogen tetroxide N_2O_4, which is changing back again at the same rate.

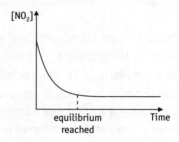

The **steady state** is important. If you started with just NO_2 then it would take some time before N_2O_4 was formed and equilibrium was established.

A dynamic equilibrium

- takes place in a *closed system* (there is no change in matter with the surroundings)
- has *forward and reverse reactions* with *equal rates*
- has *constancy of macroscopic properties* – it looks as if the system is not changing, but at the molecular level the forward and reverse reactions are occurring constantly.

Le Chatelier's principle

> **Whenever a dynamic equilibrium is disturbed, it changes so that the disturbance is opposed and equilibrium is restored.**

What does this mean? Consider the equilibrium reaction $PCl_5(g) \rightleftharpoons PCl_3(g) + Cl_2(g)$. This equilibrium is exothermic, which means that the forward reaction from left to right is exothermic (the back reaction is endothermic).

- If the temperature is raised, the equilibrium will shift to the left, which is the endothermic reaction, to try and use up the heat applied.
- If the temperature is reduced, the equilibrium will shift to the right to produce more heat and counteract the reduced temperature.
- If the pressure is increased then the volume is reduced, and the equilibrium will move to the left because this side of the reaction has only 1 mole of gas, which takes up less room.
- If the pressure is reduced then the volume is increased, and the equilibrium will move to the right because here there are 2 moles of gas, which will fill the increased volume.

You can see that there are two types of change to an equilibrium which you must be able to explain – a change in **temperature** and a change in **concentration** (aqueous reactions) **or pressure** (gaseous reactions). These changes are summarised below.

Summary of what happens when the temperature of an equilibrium changes, according to Le Chatelier

| exothermic reaction ΔH negative | temperature *increase* | reaction goes to the *left* (increases the back reaction) |
| | temperature *decrease* | reaction goes to the *right* (increases the forward reaction) |

| endothermic reaction ΔH positive | temperature *increase* | reaction goes to the *right* (increases the forward reaction) |
| | temperature *decrease* | reaction goes to the *left* (increases the back reaction) |

✓ *Quick check 2*

Summary of what happens when the concentration (or pressure) of an equilibrium changes, according to Le Chatelier

If the concentration (or pressure) of any reactant is increased, the reaction moves to the right. This uses up the extra reactant.

If the concentration (or pressure) of any product is increased, it moves to the left. This uses up the extra product.

Example: CH_3COOH (aq) + C_2H_5OH (aq) $\rightleftharpoons$ $CH_3COOC_2H_5$ (aq) + H_2O (l)

In this equilibrium, if the concentration of C_2H_5OH is increased, the reaction makes more products. If it is decreased, it makes more reactants.

Example: $2SO_2$ (g) + O_2 (g) $\rightleftharpoons$ $2SO_3$

In this reaction, there are 3 moles of gas on the left and 2 moles of gas on the right. If the pressure is *increased*, the volume is reduced so the reaction moves to the *right*. If the pressure is *decreased* the reaction moves to the *left*.

✓ *Quick check 1*

? Quick check questions

1 In the following equilibria, state the effect of increasing the pressure:
   (a) $COCl_2(g) \rightleftharpoons CO(g) + Cl_2(g)$;
   (b) $CS_2(g) + 4H_2(g) \rightleftharpoons CH_4(g) + 2H_2S(g)$;
   (c) $4NH_3(g) + 5O_2(g) \rightleftharpoons 4NO(g) + 6H_2O(g)$;
   (d) $H_2(g) + I_2(g) \rightleftharpoons 2HI(g)$

2 The gas nitrogen dioxide dimerises: $2NO_2(g) \rightleftharpoons N_2O_4(g)$ $\Delta H = -54$ kJ mol^{-1}
   Explain why the amount of nitrogen dioxide is increased when the temperature of this equilibrium is increased.

The Haber process

The Haber process is a very important industry, because it makes ammonia which is then used to make *fertilisers* (see page 80), as well as much smaller amounts of *nitric acid* and *nylon*. The nitric acid is used to make *explosives*.

N_2 (g) + $3H_2$ (g) $\rightleftharpoons$ $2NH_3$ (g)

ΔH is negative, exothermic forward reaction

This is an exothermic reaction, so decreasing the temperature will produce more ammonia. The trouble is that a low temperature *slows down* the rate of the reaction, which is not desirable in a commercial process.

This reaction will produce more ammonia if the pressure is increased, because there are only 2 moles of gas on the left and 4 moles on the right. The trouble is that doing a chemical reaction under high pressure means special equipment and safety measures, which increases the cost and so is not desirable in a commercial operation.

In real life, a *compromise* is reached which gives good yields of ammonia but keeps the costs at a manageable level. The conditions used are around 25×10^3 kPa or 250 atm of pressure and 450°C.

? Quick check questions

1 The production of ammonia gas is one of the most important chemical industries. Explain why this is so.

2 Ammonia gas is produced according to the following equation.

N_2 (g) + $3H_2$ (g) $\rightleftharpoons$ $2NH_3$ (g)

This is an exothermic reaction.

(a) Explain whether a high pressure will increase the yield of ammonia.

(b) Explain whether a high temperature will increase the yield of ammonia.

(c) State and explain the actual conditions of pressure and temperature used for the production of ammonia gas.

Acids and bases

You already know that acids have a pH below 7, and alkalis have a pH above 7. Now you have to learn exactly what an acid is.

> **An acid is a proton donor.**
> **A base is a proton acceptor.**

We can show this by looking at the equilibria which form when acids and bases dissociate in aqueous solution:

- an acid donates H^+ ions, so if HA is an acid

HA (aq) + H_2O $\rightleftharpoons$ H_3O^+ (aq) + A^- (aq)

- a base accepts H^+ ions, so if B is a base

B (aq) + H_2O (l) $\rightleftharpoons$ BH^+ (aq) + OH^- (aq)

Examples:

$$HCl \text{ (aq)} + H_2O(l) \rightleftharpoons H^+ \text{ (aq)} + Cl^- \text{ (aq)}$$

monobasic acid – donates 1 proton per molecule of acid

$$H_2SO_4 \text{ (aq)} + H_2O(l) \rightleftharpoons 2H^+ \text{ (aq)} + SO_4^- \text{ (aq)}$$

dibasic acid – donates 2 protons per molecule of acid

$$NH_3 \text{ (aq)} + H_2O \text{ (l)} \rightleftharpoons NH_4^+ \text{ (aq)} + OH^- \text{ (aq)}$$

base – accepts H^+ from H_2O

You can work out which species is the acid in a reaction by looking to see which has lost H^+.

Example: $HCl(aq) + HBr(aq) \rightarrow H_2Cl^+ \text{ (aq)} + Br^-$

Normally, you might expect HCl to be an acid, but in this particular reaction it has accepted the H^+ to become H_2Cl^+ – it is the base! HBr is the acid because it has lost H^+ to become Br^-. This is because HCl is a weaker acid than HBr so the it is forced to play the part of the base and accept the H^+.

In the following reactions, the acids are underlined:

$$\underline{HSO_4^-} \text{ (aq)} + HNO_3(aq) \rightarrow SO_4^{2-} \text{(aq)} + H_2NO_3^+\text{(aq)}$$
$$\underline{H_3PO_4}(aq) + H_2O(l) \rightarrow HPO_4^- \text{ (aq)} + H_3O^+\text{(aq)}$$

> All these equilibria lie well to the right, so we often write them as equations with an arrow.

> What's the difference between a base and an alkali? Both accept protons. A base is *any species that accepts protons,* such as solid magnesium oxide or aqueous ammonia. An alkali means *aqueous OH⁻ ions only,* such as aqueous sodium hydroxide.

> ✓ *Quick check 1*

Typical reactions of acids

Acids have four typical reactions which you must learn.
It is useful to know these as ionic equations too.

1 with reactive metals, acids give a salt + hydrogen

(reactive metals are above hydrogen in the reactivity series).

equation is $\quad Ca \text{ (s)} + H_2SO_4 \text{ (aq)} \rightarrow CaSO_4 \text{ (aq)} + H_2 \text{ (g)}$
ionic equation is $\quad Ca \text{ (s)} + 2H^+ \text{ (aq)} \rightarrow Ca^{2+} \text{ (aq)} + H_2 \text{ (g)}$

2 with carbonates, acids give a salt + carbon dioxide + water

equation is $\quad CaCO_3 \text{ (s)} + 2HCl \text{ (aq)} \rightarrow CaCl_2 \text{ (aq)} + CO_2 \text{ (g)} + H_2O \text{ (l)}$
ionic equation is $\quad CO_3^{2-} \text{ (aq)} + 2H^+ \text{ (aq)} \rightarrow CO_2 \text{ (g)} + H_2O \text{ (l)}$

Reactivity series showing some metals only	
most reactive metal	lithium
	potassium
	calcium
	sodium
	magnesium
	aluminium
	zinc
	iron
	HYDROGEN
	copper
least reactive metal	silver

3 with bases, acids give a salt – this is a neutralisation reaction

equation is $\quad$ HCl (aq) + NH$_3$ (aq) $\quad \rightarrow$ NH$_4$Cl (aq)

ionic equation is $\quad$ H$^+$ (aq) + NH$_3$ (aq) $\quad \rightarrow$ NH$_4^+$

4 with alkalis, acids give a salt + water – this is a neutralisation reaction

equation is $\quad$ H$_2$SO$_4$ (aq) + 2NaOH $\quad \rightarrow$ Na$_2$SO$_4$ (aq) + H$_2$O (l)

ionic equation is $\quad$ H$^+$ (aq) + OH$^-$ (aq) $\quad \rightarrow$ H$_2$O (l)

Strong and weak acids

We have seen that acids dissociate in solution to give H$^+$ ions

- A **strong acid** dissociates completely, so the equilibrium lies completely to the right. Hydrochloric, sulphuric and nitric acids are all strong acids:

$$\text{HCl (aq)} \rightarrow \text{H}^+ \text{ (aq)} + \text{Cl}^- \text{ (aq)}$$

- A weak acid does not dissociate completely, so in the solution there are always some complete acid molecules present, as well as H$^+$ ions. Organic acids such as ethanoic acid, are often weak acids.

$$\text{CH}_3\text{COOH (l)} + \text{H}_2\text{O (l)} \leftrightharpoons \text{CH}_3\text{COO}^- \text{ (aq)} + \text{H}_3\text{O}^+ \text{ (aq)}$$

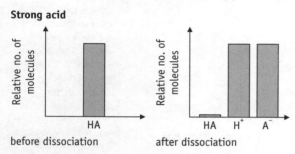

Strong acid

before dissociation $\qquad$ after dissociation

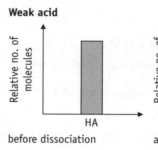 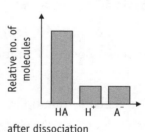

Weak acid

before dissociation $\qquad$ after dissociation

A weak acid generally has a higher pH than a strong acid. This is because, if we compare the same amounts of a weak and a strong acid, the weak acid has a lower concentration of H$^+$ ions in solution.

Ammonia as a base

The reaction of ammonia with sulphuric acid is important because the salt produced, ammonium sulphate, is used as a fertiliser. It provides nitrogen which is an important nutrient for plants.

$$\text{NH}_3 \text{ (aq)} + \text{H}_2\text{SO}_4 \text{ (aq)} \rightarrow \text{(NH}_4)_2\text{SO}_4 \text{ (aq)} + \text{H}_2\text{O (l)}$$
$$\text{ammonium sulphate}$$

Common fertilisers are ammonium sulphate, ammonium nitrate (NH$_4$NO$_3$) and urea (CO(NH$_2$)$_2$. Ammonia can also be applied directly onto the soil to act as a fertiliser.

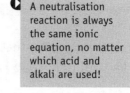

? Quick check questions

1 Identify the acid in the forward reaction of:

(a) HSO$_4^-$ + HNO$_2$ $\leftrightharpoons$ H$_2$NO$_2^+$ + SO$_4^{2-}$

(b) HCO$_3^-$ + H$_2$PO$_4^-$ $\leftrightharpoons$ HPO$_4^{2-}$ + H$_2$O + CO$_2$

2 Complete these reactions, and balance if necessary:

(a) H$_2$SO$_4$ + Ba $\rightarrow$ $\qquad\qquad$ (b) HCl + Mg(OH)$_2$ $\rightarrow$

3 Explain why the concentration of hydrogen ions is greater for a strong acid, such as sulphuric acid, than for a weak acid, such as acetic acid.

Module C: End-of-module questions

1 (a) Using the axes opposite, sketch the reaction pathway for an exothermic process. On your diagram label

   (i) the enthalpy change of reaction, ΔH,

   (ii) the activation energy of the reaction, E_a. [3]

   (b)(i) Using the axes opposite, sketch the distribution of energies of the molecules of a gas at temperature, T_1. Label this line T_1.

   (ii) On the same axes, sketch the distribution of energies of the molecules of the gas at a higher temperature T_2. Label this line T_2.

   (iii) Use your sketches to explain how a higher temperature affects the rate of a reaction. [7]

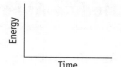

2 Methane production as 'Biogas' is growing rapidly as an alternative energy supply, particularly in some countries. Methane can be used as a fuel because of its exothermic reaction with oxygen.

$$CH_4(g) + 2O_2(g) \rightarrow CO_2(g) + 2H_2O(l) \quad \Delta H_c^{\ominus}(CH_4(g)) = -890.3 \text{ kJ mol}^{-1}$$

   (a) Explain what is meant by $\Delta H_c^{\ominus}(CH_4(g)) = -890.3$ kJ mol^{-1}. [3]

   (b) The enthalpy change of formation of methane, $\Delta H_f^{\ominus}(CH_4(g))$, cannot be measured directly.

   (i) Using the data below, calculate the enthalpy change of formation of methane. $C(s) + 2H_2(g) \rightarrow CH_4(g)$

Compound	$\Delta H_c^{\ominus}$ / kJ mol^{-1}
$CH_4(g)$	− 890.3
$C(s)$	− 393.5
$H_2(g)$	− 285.9

   (ii) Suggest why the enthalpy change of formation of methane cannot be measured directly. [4]

   (c) A typical Biogas plant in China, using the dung from five cows, can produce 3000 dm^3 of Biogas a day. The Biogas contains 60% of methane by volume.

   (i) Using the data above, calculate the maximum heat energy that can be produced each day from the methane in the Biogas. (Assume 1 mole of methane occupies 24 dm^3 under these conditions)

   (ii) Suggest a practical difficulty of using Biogas on a large scale. [4]

Answers to quick check questions

Module A: Foundation Chemistry

Atomic structure Page 3

1 $^{40}_{20}$Ca: 20p, 20e, 20n $^{16}_{8}$O: 8p, 8e, 8n

$^{14}_{6}$C: 6p, 6e, 8n $^{12}_{6}$C: 6p, 6e, 6n

$^{19}_{9}$F$^-$: 9p, 10e, 10n $^{27}_{13}$Al^{3+}: 13p, 10e, 14n

2 $Ca(OH)_2$ $Pb_3(PO_4)_2$ $BaCO_3$ K_2SO_4

3 $N_2 + 3H_2 \rightarrow 2NH_3$

$3Fe + 4H_2O \rightarrow Fe_3O_4 + 4H_2$

$Na_2O + H_2O \rightarrow 2NaOH$

$PCl_5 + 4H_2O \rightarrow H_3PO_4 + 5HCl$

4 (a) $Mg(s) + H_2SO_4(aq) \rightarrow MgSO_4(aq) + H_2(g)$

(b) $Mg(s) + 2H^+(aq) \rightarrow Mg^{2+}(aq) + H_2(g)$

5 44

6 132

Mass spectrometer Page 5

1

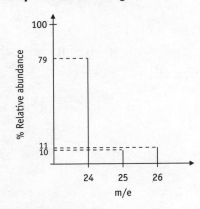

2 (a) 107.9 (b) 35.49

Mole calculations Page 9

1 6×10^{23} atoms

2 32g

3 0.17 mol

4 (a) 1.34 g (b) 1.39 g

Empirical formulae Page 11

1 Empirical formula $C_6H_{11}O_2$; molecular formula $C_{12}H_{22}O_4$

2 C_2H_6O

Titration calculations Page 12

1 0.06 mol dm^{-3}

2 50 cm^3

Electron configurations Page 13

1 V: $1s^2 2s^2 2p^6 3s^2 3p^6 4s^2 3d^3$

K$^+$: $1s^2 2s^2 2p^6 3s^2 3p^6$

O^{2-}: $1s^2 2s^2 2p^6$

2 Magnesium

Chemical bonding Page 15

1 (a)+(b)

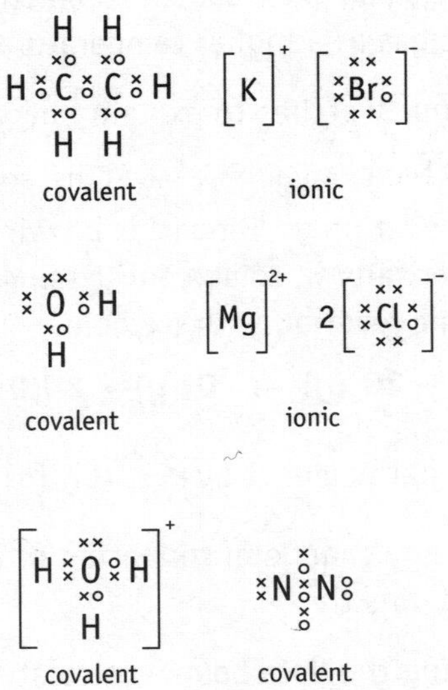

covalent   ionic

covalent   ionic

covalent   covalent

Polarisation Page 16

1 H—I Yes; Cl—P—Cl Yes; CH_4 No

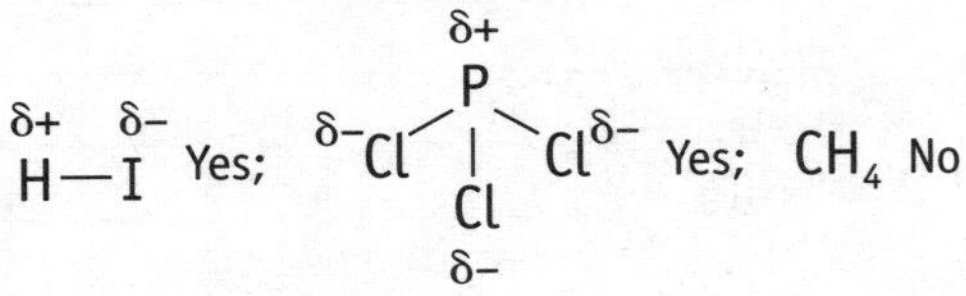

Intermolecular forces Page 17

1 C_3H_8 v.d. Waals' forces; NH_2OH H-bonding; PCl_3 dipole-dipole attractions

2

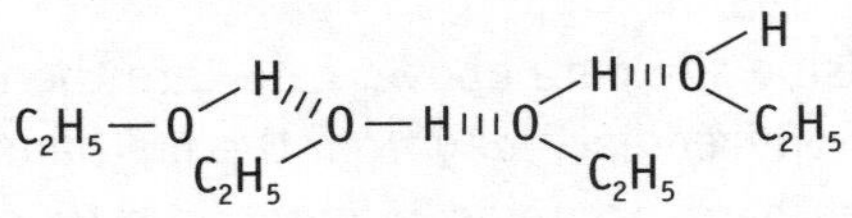

3 Methoxymethane cannot form H-bonds, ethanol can. To break H-bonds requires additional heat.

Structure and properties Page 19

1 (a) Giant ionic lattice (b) Metallic (c) Discrete atoms (d) Discrete molecules

2 Giant covalent structure – v. high m.p. and v. hard indicates covalent bonds in a network.

Shapes of molecules Page 21

1 ClF linear; PCl_3 bond angle 107°; OCl_2 bond angle 104°; XeF_6 octahedral, bond angle 90°; CCl_4 tetrahedral, bond angle 109.5°.

Periodicity Page 23

1 All the elements in the p-block have outermost electrons in a p sub-shell.

2 A $[Kr]5s^2$; B $[He]2s^22p^1$; C $[Ne]3s^1$; D $[He]2s^22p^6$

Periodicity Page 25

1 Al, Na, P. Al and Na are metals so they conduct. Al is the better conductor as three electrons are delocalised, increasing the electrostatic attraction of the metallic bond. P is a non-metal, with a covalent structure and therefore no free electrons, so does not conduct.

2 In Na **one electron only** is lost to the metallic bond, so Na has the lowest electrostatic attraction holding the structure together.

Ionisation energies Page 27

1

2 see text

3 approx. 940 $kJ\,mol^{-1}$

Group 2 elements Page 29

1 Increases

2 $BaSO_4$ used in barium meals – see text

3 $Ca(OH)_2(aq) + CO_2(g) \rightarrow CaCO_3(s) + H_2O(l)$

Oxidation and reduction Page 31

1 Reduction is:
the gain of hydrogen (e.g. $Li + \frac{1}{2}H_2 \rightarrow LiH$);
the loss of oxygen (e.g. $PbO_2 \rightarrow PbO + \frac{1}{2}O_2$);
the gain of electrons (e.g. $\frac{1}{2}Cl_2 + e^- \rightarrow Cl^-$);
a decrease in oxidation number
(e.g. $\frac{1}{2}Cl_2 + e^- \rightarrow Cl^-$ ON of $Cl_2 = 0$, ON of $Cl^- = -1$).

2 (a) +4 (b) −3 (c) +4

3 **NaCl** Na +1, Cl−1; **LiH** Li+1, H−1;
MgBr₂ Mg +2, Br −1; **NaOH** Na +1, O −2, H +1,
AgNO₃ Ag +1, N +5, O −2; **Na₂SO₄** Na +1,
S +6, O −2; **Cl₂** Cl 0.

4 (a) oxidised (0 → +2)
(b) oxidised (+2 → +2.5)
(c) reduced (+3 → −2)

Reactions of Group 2 metals Page 33

1 Mg(0)→Mg(II), oxidised

2 Effervescence – H_2 gas released, Ca disappears

3 0.24 dm^3

Group 7 elements Page 35

1 $Br_2(aq) + 2I^-(aq) \rightarrow 2Br^-(aq) + I_2(aq)$ cyclohexane turns orange-brown.

2 I_2 is the larger molecule so has greater van der Waals' forces

Halogens Page 37

1 (a) +5 (b) $Cl_2(0) \rightarrow Cl^- (-1) + ClO_3^- (+5)$
Cl is reduced and oxidised in the same reaction.

2 ClO^- +1; Cl^- −1; ClO_3^- +5

3 see text.

Module B: Chains and Rings

Formulae and isomerism Page 41

1 (a)

Pentane   2-Methylbutane

2,2 -Dimethylpropane

(b)

Propan-2-ol   Propan-1-ol

2 Positional isomerism because the position of the OH group can change.

3 (a)

(b)

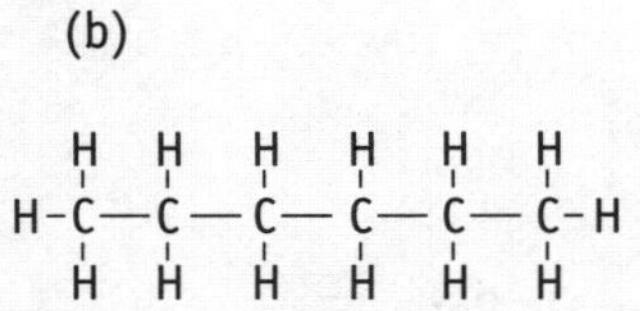

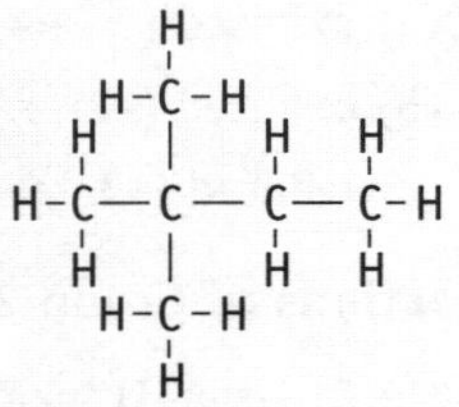

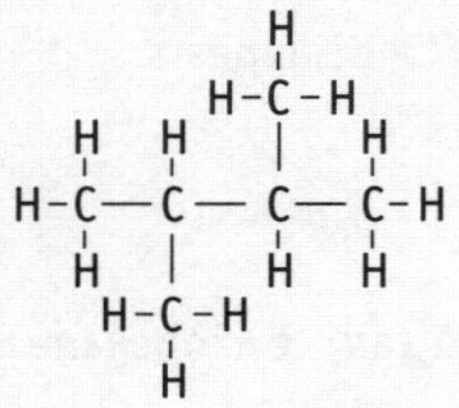

4 $CH_3CH_2CH_2CH_2OH$ Butan-1-ol $CH_3CHCH_2CH_3$ with OH Butan-2-ol

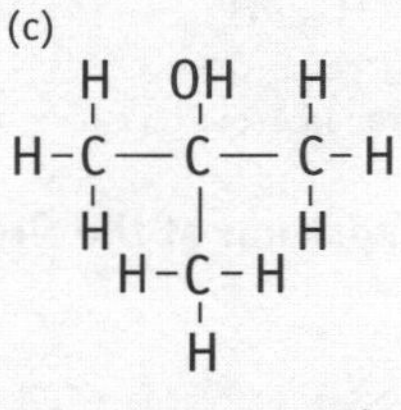

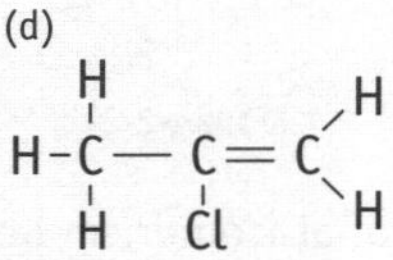

2-Methylpropan-2-ol   2-Methylpropan-1-ol

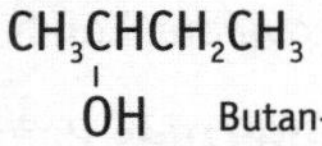

The alkanes Page 45

1 C_8H_{18}

2 $C_5H_{12} + 8O_2 \rightarrow 5CO_2 + 6H_2O$

3 (a)

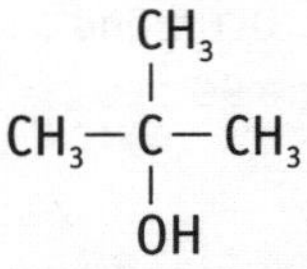

Butane

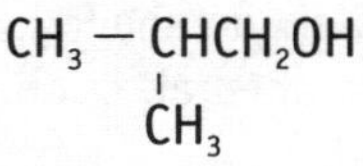

2-Methylpropane

(b) (i) Butane ($CH_3CH_2CH_2CH_3$)

   (ii) It has more points of contact (larger surface area) and therefore the van der Waal's forces between the molecules are stronger.

4 Two $CH_3 \bullet$ (methyl) radicals combine to form ethane.

   $CH_3 \bullet + CH_3 \bullet \rightarrow H_3C{:}CH_3$ (C_2H_6)

4

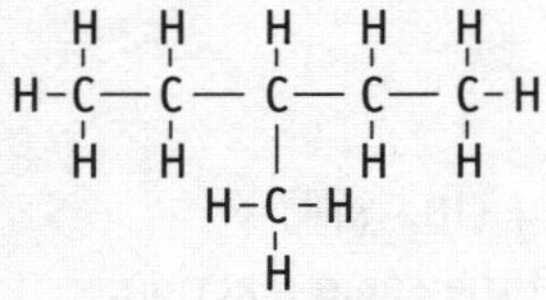

Pent-1-ene 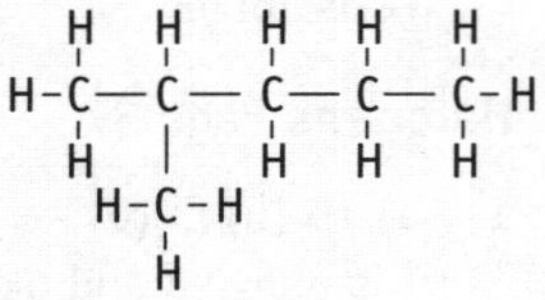   3-Methylbut-1-ene

It is O.K. in many displayed formulae to show methyl groups as $-CH_3$

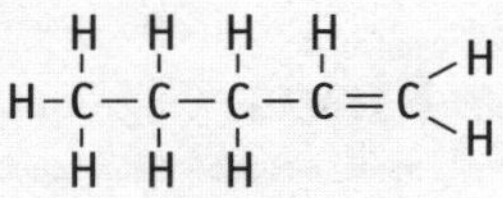

Cis-pent-2-ene 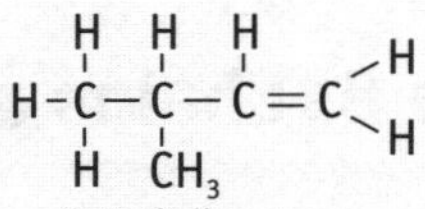   Trans-pent-2-ene 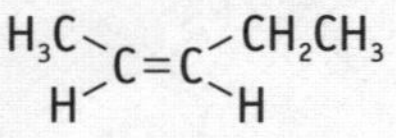   2-Methylbut-1-ene

5 Functional group isomerism because they have different functional groups. Ethanol has the OH group and is an alcohol. Diethyl ether has the ether C–O–C group.

Naming organic compounds Page 43

1 (a) butan-2-ol (b) 1-chloropropane
   (c) but-2-ene (d) 2-bromopropane

2

(a)

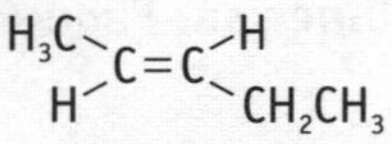

(b)

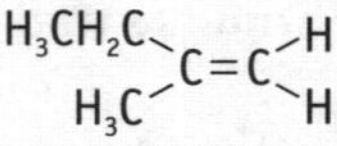

(c)

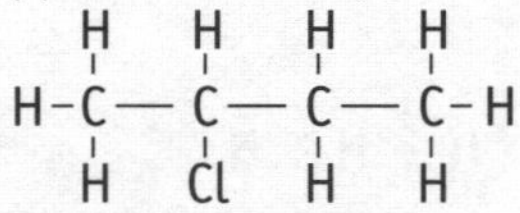

(d)

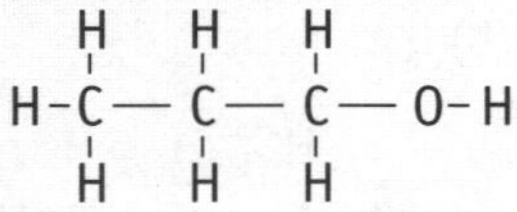

3 You must number the carbon chain from the lowest possible carbon for the functional group. In these cases it **can only be** carbon number 1.

Hydrocarbons and fuels Page 47

1 $C_7H_{16} + 11O_2 \rightarrow 7CO_2 + 8H_2O$. The products are both gases at the very high temperatures of the car engine and these occupy a greater volume than the reactants.

2 Octane into 2,2,4-trimethylpentane is an example of isomerisation (see text for details).

   Octane into 1,4 dimethylbenzene is an example of reforming (see text for details).

3 (a) (i) $C_{12}H_{26} \rightarrow C_{10}H_{22} + C_2H_4$

      (ii) $C_{12}H_{26} \rightarrow C_8H_{18} + C_4H_8$

   (b) (i) They ignite prematurely and cause knocking (see text for details).

      (ii) Cyclisation and/or reforming. (see text for details).

The alkenes Page 49

1

$$H-\underset{\underset{H}{|}}{\overset{\overset{H}{|}}{C}}-\underset{\underset{H}{|}}{\overset{\overset{H}{|}}{C}}-C=C\overset{H}{\underset{H}{}} \qquad \overset{H_3C}{\underset{H_3C}{}}C=C\overset{H}{\underset{H}{}}$$

$$\overset{H}{\underset{H_3C}{}}C=C\overset{CH_3}{\underset{H}{}} \qquad \overset{H_3C}{\underset{H}{}}C=C\overset{CH_3}{\underset{H}{}}$$

2 The numbers of pairs of electrons round each carbon is four. One of these pairs is used in a Π bond and does not contribute to the shape. Therefore the real number of electrons pairs is three and the shape is trigonal planar with a bond angle of $120^\circ C$.

3 (a) *Cis-trans* or geometric isomerism.

(b) The structural formulae are identical but the displayed formulae showing the arrangements of the atoms and bonds in space are different.

(c) (i) No. There are two identical groups on the same carbon.

(ii) Yes. The CH_3 and the C_2H_5 groups are on different carbons and therefore they can be on the same or opposite sides of the C=C bond.

The alkenes Page 51

1 (a) (i) Conditions = Phosphoric acid (H_3PO_4) catalyst; $300^\circ C$ and 70 atmospheres pressure.

Product = ethanol, CH_3CH_2OH.

(ii) Conditions = Nickel catalyst; $400\,^\circ C$.

Product = ethane (CH_3CH_3)

(iii) Conditions = Room temperature with HBr.

Product = bromoethane, CH_3CH_2Br.

(iv) Conditions = Br_2 in hexane at room temperature (in the dark). Bromine water would be acceptable).

Product = 1,2–dibromoethane, CH_2BrCH_2Br

(b) (i) Conditions as for (a) (i). Product = butan-2-ol, $CH_3CH(OH)CH_2CH_3$

(ii) Conditions as for (a) (ii). Product = butane, $CH_3CH_2CH_2CH_3$

(iii) Conditions as (a) (iii). Product = 2-bromobutane, $CH_3CHBrCH_2CH_3$.

(iv) Conditions as for (a) (iv). Product = 2,3-dibromobutane, $CH_3CHBrCHBrCH_3$.

2 (a) CH_2 (b) C_3H_6;

$$H-\underset{\underset{H}{|}}{\overset{\overset{H}{|}}{C}}-\underset{\underset{H}{|}}{\overset{\overset{H}{|}}{C}}=C\overset{H}{\underset{H}{}}$$

(c)

$$H-\underset{\underset{H}{|}}{\overset{\overset{H}{|}}{C}}-\underset{\underset{H}{|}}{\overset{\overset{OH}{|}}{C}}-\underset{\underset{H}{|}}{\overset{\overset{H}{|}}{C}}-H$$

$$H-\underset{\underset{H}{|}}{\overset{\overset{H}{|}}{C}}-\underset{\underset{H}{|}}{\overset{\overset{H}{|}}{C}}-\underset{\underset{H}{|}}{\overset{\overset{H}{|}}{C}}-OH$$

(d)

$$H-\underset{\underset{H}{|}}{\overset{\overset{H}{|}}{C}}-\underset{\underset{Br}{|}}{\overset{\overset{H}{|}}{C}}-\underset{\underset{H}{|}}{\overset{\overset{H}{|}}{C}}-H$$

$$H-\underset{\underset{H}{|}}{\overset{\overset{H}{|}}{C}}-\underset{\underset{H}{|}}{\overset{\overset{H}{|}}{C}}-\underset{\underset{H}{|}}{\overset{\overset{H}{|}}{C}}-Br$$

3 See text for answer.

4 Margarine is produced from oils which have unsaturated chains (contain C=C bonds) on their molecules. Hydrogenation of these oils using a nickel catalyst gives saturated chains (with no double bonds). Animal fats have chains which are mostly already saturated and therefore have no C=C bonds which can react with hydrogen.

Addition polymerisation Page 53

1 (a) (i)
$$\overset{H}{\underset{H}{}}C=C\overset{H}{\underset{H}{}}$$

(ii)
$$\overset{H_3C}{\underset{H}{}}C=C\overset{H}{\underset{H}{}}$$

(iii)
$$\overset{H}{\underset{H}{}}C=C\overset{Cl}{\underset{H}{}}$$

(iv)
$$\overset{F}{\underset{F}{}}C=C\overset{F}{\underset{F}{}}$$

(b) (i)
$$\cdots\overset{}{\underset{}{C}}-\overset{}{\underset{}{C}}-\overset{}{\underset{}{C}}-\overset{}{\underset{}{C}}\cdots$$ (with H H H H top and H H H H bottom)

(ii)
$$\cdots\overset{}{\underset{}{C}}-\overset{}{\underset{}{C}}-\overset{}{\underset{}{C}}-\overset{}{\underset{}{C}}\cdots$$ (with CH_3 H CH_3 H top and H H H H bottom)

(iii)
$$\cdots\overset{}{\underset{}{C}}-\overset{}{\underset{}{C}}-\overset{}{\underset{}{C}}-\overset{}{\underset{}{C}}\cdots$$ (with H Cl H Cl top and H H H H bottom)

(iv)
$$\cdots\overset{}{\underset{}{C}}-\overset{}{\underset{}{C}}-\overset{}{\underset{}{C}}-\overset{}{\underset{}{C}}\cdots$$ (with F F F F top and F F F F bottom)

2 (a) Carbon monoxide

(b) Carbon monoxide, hydrogen chloride and polychlorinated biphenyls.

3 (a)
$$\overset{NC}{\underset{H}{}}C=C\overset{CH_3}{\underset{H}{}}$$

(b) Carbon monoxide, oxides of nitrogen (NO_x) or more specifically nitrogen dioxide, hydrogen cyanide.

The alcohols Page 55

1   (a) But-1-ene, $CH_3CH_2CH=CH_2$.

  (b) Sodium butoxide, $CH_3CH_2CH_2CH_2O^-Na^+$ and hydrogen (H_2)

  (d) 2-bromo-2-methylpropane, $CH_3C(CH_3)BrCH_3$ and water.

2   (a)   (b)

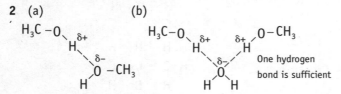

One hydrogen bond is sufficient

3   (a) $C_2H_5OH + 3O_2 \rightarrow 2CO_2 + 3H_2O$

  (b) (i) It is easily available (from fermentation of sugars or hydration of ethene).

  It is easily transported.   It ignites easily.

  When it burns a considerable amount of heat is given off.

  It burns 'cleanly' to give non-polluting products (although CO_2 contributes to global warming).

  When it burns the volume of the products is greater than the reactants.

  (ii)   Ethanol can be produced from sugar by fermenation. Sugar cane can be grown again and again, will not run out and is therefore renewable.

The alcohols Page 57

1   (a) Primary   (b) Secondary   (c) Primary
  (d) Tertiary   (e) Secondary

2   Use acidified potassium dichromate solution.

  This is orange in colour.

  Heat it separately with the two alcohols.

  The butan-2-ol is oxidised by the acidified potassium dichromate to a butan-2-one (2° alcohols are oxidised to ketones) and therefore the orange colour changes to a blue-green colour as the potassium dichromate is reduced.

  The 2-methylpropan-2-ol is unaffected since it is a tertiary alcohol and cannot be oxidised under these conditions, therefore the orange colour is unchanged.

3   Once the percentage of alcohol reaches 12% the yeast cells are killed and the fermentation cannot proceed any further. Therefore the alcohol can never reach 100% by this method.

4   In the presence of air (oxygen) and bacteria (which supply the enzymes), the ethanol (a primary alcohol) can be oxidised to an aldehyde and then a carboxylic acid. The carboxylic acid in this case is ethanoic acid. Vinegar is a dilute solution of ethanoic acid.

Infrared spectroscopy Page 58

1   (a) OH group
  (b) alcohol or phenol

2   At first the absorption due to the OH group (at ~ 3400 cm^{-1} the alcohol will disappear as this is oxidised to a carbonyl (>C=O) group in the aldehyde. At the same time the absorption at ~1700 cm^{-1} appears and gets stronger.

  As the aldehyde is then oxidised to a carboxylic acid a broad absorption at 2500–3300 cm^{-1} appears. The absorption due to the carbonyl group remains.

3   The ethoxyethane has no OH group and therefore there will be no absorption at ~3400 cm^{-1}.

4   The propan-2-ol will not absorb at ~1700 cm^{-1} because it has no carbonyl (>C=O) group.

The halogenoalkanes Page 61

1   (a) C_3H_7Br

  (b) C_3H_7Br

  (c)

  H H H   H H H
  | | | | | |
H–C — C — C–H   H–C — C — C–Br
  | | | | | |
  H Br H   H H H

2   The isomers are:

  $CH_3CH_2CH_2CH_2Br$ 1-bromobutane   primary

  $CH_3CHBrCH_2CH_3$ 2-bromobutane   secondary

  $CH_3CBr(CH_3)CH_3$ 2-bromo-2-methylpropane
                                                  tertiary

3   (a) Ammonia is a polar molecule (with the nitrogen having a partial negative charge because of its higher electronegativity) and has a lone pair on the nitrogen ($:NH_3$). Therefore it is firstly attracted to the electron-deficient carbon in the carbon-halogen bond and has a lone pair of electrons which it can donate to the carbon during the reaction.

  (b) KCN contains the $CN:^-$ ion which is negative and is therefore attracted to the electron-deficient carbon in the carbon-halogen bond and has a lone pair of electrons which it can donate to the carbon during the reaction.

  (c) KOH contains the OH^- ion which is a nucleophile for the same reasons as the CN^- ion.

4   (a) The synthesis relies on the halogen being replaced by another group and therefore the breaking of the carbon–halogen bond. The C–F bond is very strong and therefore difficult to break so the reaction would be extremely slow and not viable as part of a synthetic route.

  (b) The C–I bond is too weak and consequently iodoalkanes often react with contaminants before it can be used in a reaction. Bromoalkanes are sufficiently reactive to make them useful in laboratory preparations whilst not so reactive that they react with contaminants before they can be used in the laboratory.

The halogenoalkanes Page 63

1 but-1-ene ($CH_3CH_2CH=CH_2$); but-2-ene ($CH_3CH=CHCH_3$) and there are two isomers of this – the *cis* and the *trans* isomers, giving three products in total.

2 (a) $CH_2BrCH_2Br + 2KOH \rightarrow 2KBr + CH_2OHCH_2OH$

 (b) $CH_2BrCH_2Br + 2KCN \rightarrow 2KBr + CH_2CNCH_2CN$

 (c) $CH_2BrCH_2Br + 2NH_3 \rightarrow 2HBr + CH_2NH_2CH_2NH_2$

3 STAGE I

Heat bromoethane with a concentrated solution of KOH in alcohol. HBr is eliminated to give ethene.

$CH_3CH_2Br + KOH \rightarrow CH_2=CH_2 + H_2O + KBr$

STAGE II

This is then polymerised in the presence of a catalyst, heat and a high pressure to give polyethene.

$n[CH_2=CH_2] \rightarrow -[CH_2-CH_2]_n-$

4 See text for details.

5 Comparable halogenoalkanes such as 1-bromobutane and 1-chlorobutane (both with 4 carbons and both primary halogenoalkanes) are used to make it a fair test.

Silver nitrate solution is dissolved in alcohol and the solution warmed gently in two separate test tubes.

Equal amounts of the 1-bromobutane and 1-chlorobutane are added separately to each tube and the appearance of a precipitate indicates the formation of insoluble silver halide.

The appearance of the precipitate indicates the breaking of the carbon-halogen bond – it is quicker for the 1-bromobutane indicating that the carbon-bromine bond is the weaker of the two.

Yields Page 64

1 18.0 g of propan-1-ol = 18/60 mol = 0.3 mol

∴0.3 mol of propanal should be obtained = 0.3 x 58 g = 17.4 g.

∴Percentage yield = (7.2/17.4) x 100% = 41.4%

2 9.2 g of ethanol is 0.2 mol (9.2/46)

∴0.2 mol of bromoethane are produced if 100% conversion takes place.

∴0.2 x 109 g (M_r of bromoethane is 109) = 21.8 g of bromoethane are produced if 100% conversion takes place.

∴percentage yield = 5.36/21.8 = 24.6%

Module C: How Far? How Fast?

Enthalpy changes Page 69

1 +26 kJ mol^{-1}

Enthalpy change definitions Page 72

1 – 75.6 kJ mol^{-1}

2 – 43.1 kJ mol^{-1}

Reaction Rates Page 74

1

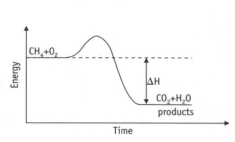

2 Homogeneous

3

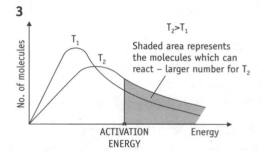

Pollution Page 75

1 See text.

Chemical Equilibrium Page 77

1 (a) Equilibrium shifts to the left

 (b) Equilibrium shifts to the right

 (c) Equilibrium shifts to the left

 (d) No change.

2 Exothermic reaction, so the back reaction is favoured when temperature increases

Haber process Page 78

1 see text.

2 see text.

Acids and bases Page 80

1 (a) HSO_4^- (b) $H_2PO_4^-$

2 (a) $H_2SO_4 + Ba \rightarrow BaSO_4 + H_2$

 (b) $2HCl + Mg(OH)_2 \rightarrow MgCl_2 + 2H_2O$

3 Strong acids dissociate completely, weak acids do not.

Answers to End-of-module questions

NOTE * indicates a mark awarded.

Module A: Foundation Chemistry

1 (a)(i) 14* (ii) 10* (iii) 8* (iv) 6*
   (b)(i) 2 mol Ca^{2+} 2 mol C^{2-}*
      (ii) +5* (iii) 32 *
   (c)(i) 69.80* (ii) 3 g * (iii) 4 g *
2 (a)(i)

$$H \overset{\times\times}{\underset{\times\times}{\circ}} \overset{\times\times}{\underset{\times}{F}} \times \quad \left[Ca \right]^{2+} 2 \left[\overset{\times\times}{\underset{\circ}{\underset{\times\times}{F}}} \times \right]^{-} **$$

   (ii) HF is a gas, CaF_2 is a solid*
         HF is discrete molecules,
         CaF_2 is an ionic lattice*
3 (a) +4 in SO_2 * and +6 in SO_4^{2-}*.
   (b)(i) 0.000164 mol * (ii) 0.000164 mol *
      (iii) 0.0032 mol dm^{-3} *
      (iv) M_R of SO_2 = 64 *, 0.2048 g * dm^{-3}

Module B: Chains and Rings

1 (a)(i) Alkenes*
      (ii) B and C*
   (b)(i) D *
      (ii) Pass vapour over heated pumice** or heat
           with concentrated sulphuric acid. **

2

alcohol	observation	organic product (if any)
butan-1-ol	orange to green*	butanal* or butanoic acid
butan-2-al	orange to green*	butan-2-one*
2-methylpropan-2-ol	no change (stays orange)	no reaction*

3 (a)*

   (b)It is non-biodegradable and will persist in
      the environment.*
      If it is burned it will give toxic products. *

4 (a) Removal of water from a compound. *
      Example is dehydration of an alcohol to give
      an alkene. * Heat with concentrated H_2SO_4
      OR pass over heated pumice. *
      Equation $C_2H_5OH \rightarrow C_2H_4 + H_2O$ *
   (b)Removal of a small molecule like HBr from a
      molecule. *
      Example is elimination of HBr from
      bromoethane to give ethene. The
      bromoethane is heated * with a
      concentrated solution of KOH in alcohol. *
      Equation $CH_3CH_2Br + KOH \rightarrow C_2H_4 + H_2O$
      $+ KBr$ *
   (c)A halogen atom is replaced by another group
      or atom. *
      Example is the substitution of Br in
      bromoethane by OH to give ethanol. *
      The bromoethane is refluxed with an aqueous
      solution of KOH. *
      $CH_3CH_2Br + KOH \rightarrow CH_3CH_2OH + KBr$ *

5 (a)E is an alkane because its formula fits the
      general formula for alkanes (C_nH_{2n+2}) *
      It is hexane. *
      The unsaturated hydrocarbon, G, is an
      alkene, and its molecular mass corresponds
      to propene, C_3H_6 (3x12 + 6 = 42). *
      ∴ F is propane (C_3H_8) *
      $C_6H_{14} \rightarrow C_3H_8 + C_3H_6$ *
      In the presence of UV light E would undergo
      substitution to give H, $C_6H_{15}Br$, bromohexane *,
      and HBr – the acidic gas I. *
      $C_6H_{14} + Br_2 \rightarrow C_6H_{15}Br + HBr$ *
      G, propene, reacts with HBr to give two
      products, 1-bromopropane and
      2-bromopropane.*
      $C_3H_6 + HBr \rightarrow C_3H_7Br$ *
   (b)Straight chain compounds produce knocking
      when they burn (see text).
      Benzene* and cyclohexane.*
   (c)

6 (a)(i) Compounds with the same molecular formula but with different structural formulae. *

(ii) Any alkane with 6 to 10 carbons e.g. C_8H_{18}. *

(b)(i) UV light. *

(ii) C_4H_9Cl. * The chlorine accounts for 35.5 of the molecular mass * leaving 57 for C_4H_9-.*

(iii) ** Either

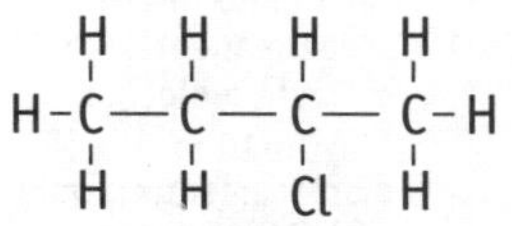

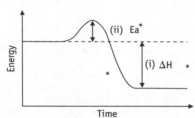

2-Chlorobutane

or

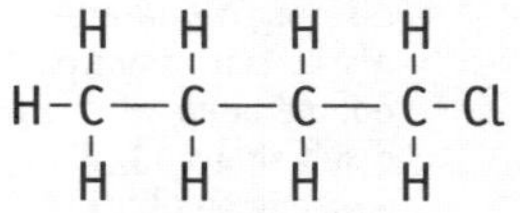

1-Chlorobutane

Module C: How Far? How Fast?

1 (a)(i) and (ii)

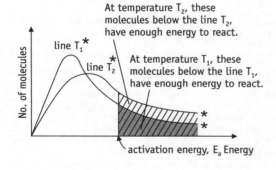

1 (b)(i) and (ii)

(iii) At the higher temperature, more molecules can react*, therefore the rate of reaction increases*.

2 (a) enthalpy change of complete combustion* of 1 mole of methane gas* under standard conditions (298K, 101 kPa)*

(b)(i)

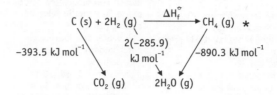

$\Delta H_f^{\ominus} = -393.5 + 2(-285.9) - (-890.3)$ *
$= -75$ kJ mol^{-1} *

(b)(ii) practical difficulty of reacting carbon with hydrogen – it does not happen at room temperature so must be heated, which is dangerous.*

(c)(i) 3000 dm^3 × 60% = 1800 dm^3*

1800 dm^3 = 75 mol CH_4*

heat energy produced = 75 × (- 890.3)
= - 6.677 kJ mol^{-1}*

(ii) problems associated with collecting the gas; large herds of cows needed etc.*

INDEX